Collaborative Writing in Industry: Investigations in Theory and Practice

Edited by
Mary M. Lay and William M. Karis

Baywood's Technical Communications Series
Series Editor: JAY R. GOULD

Baywood Publishing Company, Inc.
Amityville, N.Y. 11701

Library of Congress Catalog Number: 90-41323
ISBN: 0-89503-071-3 (cloth)
ISBN: 0-89503-070-5 (paper)

Library of Congress Cataloging-in-Publication Data

Collaborative writing in industry : investigations in theory and
    practice / edited by Mary M. Lay and William M. Karis.
        p.   cm. — (Baywood's technical communications series)
    Includes bibliographical references and index.
    ISBN 0-89503-071-3 (cloth). — ISBN 0-89503-070-5 (pbk.)
    1. Technical writing. 2. Authorship—Collaboration. I. Lay,
    Mary M. II. Karis, William M. II. Series: Baywood's technical
    communications series (Unnumbered)
    T11.C562   1991
    808'.0666—dc20   90-41323
                                                               CIP

Table of Contents

3

Introduction

MARY M. LAY
WILLIAM M. KARIS

The process by which this book has taken shape in some ways mirrors the very process it attempts to describe. This phenomenon is reflected in two aspects of the book's genesis: first, since we were originally influenced by other researchers in the field to begin this project, we are now completing that cycle by contributing to the ongoing scholarly dialogue concerning collaboration; second, this contribution to the field results from the co-editors, the contributors, and the reviewers having collaborated in many different ways.

As we originally attempted to define collaborative writing in industry we chose to frame it as "that process in which authors with differing expertise and responsibilities interact during the invention and revision of a common document." We suspected, however, and confirmed during the process of completing this project, that our notion of "authors" (or "players") must expand to include editors, graphics experts, and even users. The collaborative "players" in our own project initially included the researchers whose work we read, including those studying how the workplace should influence the classroom. For example, Ede and Lunsford called for other teachers and researchers to address "the dichotomy between current models and methods of teaching writing, almost all of which assume single authorship, and the actual writing situations students will face upon graduation, many of which require co- or group authorship" [1, p. 69]. And Jone Rymer (Goldstein) noted that collaborative writing assignments can assist students in learning to appreciate the "complex constraints" that writing on the job might entail: for example, "team writing, writing for review, writing for someone else's signature, group planning with individual drafting . . ." [2, p. 27].

Simultaneously, we also reviewed the work of those researchers describing collaborative writing within industry and calling for recognition of its pervasive nature and methods to make it more efficient. Both Steve Doheny-Farina and Mary Beth Debs, who have done ethnographic studies of collaboration in industry, for example, helped us discover the inescapable need for improved interpersonal skills if collaborative efforts are to be streamlined and made more effective [3, 4].

Thus our entry to the dialogue began with the work of those who looked at the workplace and the classroom and those who saw potential connections between. Eventually we came to appreciate Anne Ruggles Gere's assessment that when "a practice merits thoughtful criticism, as well as generalized enthusiasm, it has come of age" [5, p. 31]. We hope that this book's publication provides one additional sign that studies on collaborative writing have reached maturation.

By means of a call for proposals, we invited other "players" to join our collaboration act. Of the many fine proposals which were submitted, we chose those which would reflect a balance between theory and practice, the workplace and the classroom. Some of the players collaborate with one another and thus there are some dual- and triple-authored pieces within this collection. Our own dual-authoring experience included our bibliography published in the *Kentucky English Bulletin* which served as our opening response to the broader dialogue and provided us with a foundation with which to read our contributors' pieces [6].

Of course, our own primary collaboration was as editors. We collaborated on the first two (out of three) stages of the editing process. We read each manuscript, took notes, then negotiated the comments and suggestions we had for the authors. In the second stage of editing, we merged our suggestions with the Series Editor's and our reviewers' to send to the authors. By this point in our collaboration, we had discovered each other to have different, though complementary, editorial tendencies. While one of us would focus on thematic issues and concerns, the other tended to be more oriented to internal logic and consistency. After the first two stages of editing, we found that we each had learned the other's tendencies and were able to anticipate each other's comments and observations. Thus, in the final stage of editing, we chose to each independently review two sections of the collection and trust each other's editing style.

We then collaborated in two additional ways: we each wrote drafts of section introductions and edited each other's work; and we collaborated on the writing of this introduction by sitting in front of a computer monitor and framing our thoughts and choosing our words together. During this collaborative process, we always envisioned our readers to be members of one or more of the following communities:

1. those who teach technical communication and use collaborative writing in their classroom assignments;
2. those who are technical communicators within industry and are required to collaborate and wish to improve their collaborative techniques;
3. those who research technical communication and wish to know what theories inform collaboration;

4. those from other disciplines who have contributed to our knowledge of collaboration and wish now to learn what we have discovered about collaboration where it is practiced frequently—the workplace;
5. those graduate students in composition, rhetoric, or technical communication who need to know the theoretical and research foundations of this topic.

Our collection offers these audiences descriptions of updated industrial practices, ideas for collaborative exercises in the classroom, and overviews of the theories which inform collaboration.

Perhaps our primary goal, then, is to assist our first two audiences to better understand each other's concerns and needs by discovering ways to make it easier for writers of all types, whether undergraduates or corporate employees, to collaborate more effectively. As Van Pelt and Gillam argue in their chapter in this book, while collaboration in industry aims at a timely and usable product, collaborative activities in the classroom should focus on the intellectual and interpersonal development of students. These objectives should be recognized as complementary, not contradictory. Indeed since most of our students are going to wind up with industry eventually, these goals can and should be seen as mutually reinforcing. Our pedagogical goals are to prepare students by offering them opportunities to *develop* collaborative skills for on-the-job tasks.

As readers work through the collection, we suspect that they will discover many exciting things, as we did during our own collaboration and while studying the research of our contributors. First, while collaborative activities of all types can be difficult and time-consuming, they have a multitude of potential benefits both direct and indirect, for both product and participant. Direct benefits of collaborative writing can include a carefully designed document, a result of what Collins and Guetzkow call the "assembly effect," whereby the group solution or product is superior to that of even the best member's individual efforts [7]. Indirect benefits may take the form of improved interpersonal skills and reading ability. Second, readers should become aware that the computer is not only a writing tool for individuals, but also provides an effective medium for collaboration. Third, we discovered that current research is moving beyond rough portrayals of collaborators to look at some of their finer characteristics such as gender, experience, age, organizational role, and even language patterns and non-verbal signals. Fourth, it is clear that one of the key concerns in the future will be how to better manage the collaborative process, including such issues as leadership and task definition. Finally, scholars in other disciplines will continue to inform our studies of collaboration, but are now able to begin gathering our varied perspectives on such topics as insight, dialogue, and software development to inform their own investigations of collaboration.

We'd like to thank all the "players" in this collaboration. We appreciate the efforts and contributions of Jay Gould, Series Editor, of Norman Cohen, our publisher, and of our anonymous reviewers. We greatly appreciate the scholarly achievements of our contributors who, collaboratively, have done so much to create this work. Our thanks also go to Judy Grant, our departmental secretary, who was always able to find whatever one of us had misplaced. We would like to thank our Clarkson students and colleagues who offered encouragement and stimulated us as we attempted to understand the collaborative process and the theories behind it. We must thank Owen Brady, Dean of Liberal Studies, who provided much of the internal support to fund this project. Finally, we'd like to thank our families and friends for sustaining us throughout this project.

In a forward looking note, we'd like to express our appreciation to Myers Corners Laboratory of IBM which has provided us with a generous funding source with which to continue our research into the principles for guiding and managing collaborative writing projects and teams.

REFERENCES

1. L. Ede and A. Lunsford, Research in Collaborative Writing, *Technical Communication, 32*:4, pp. 69–70, 1985.
2. J. Rymer [Goldstein], Trends in Teaching Technical Writing, *Technical Communication, 31*:4, pp. 25–34, 1984.
3. S. Doheny-Farina, Writing in an Emerging Organization: An Ethnographic Study, *Written Communication, 3*:2, pp. 158–185, 1986.
4. M. Debs, Collaborative Writing: A Study of Technical Writing in the Computer Industry, Dissertation, Rensselaer Polytechnic Institute, 1986.
5. A. Ruggles Gere, *Writing Groups: History, Theory, and Implications,* Southern Illinois University Press, Carbondale, p. 31, 1987.
6. W. Karis and M. Lay, Collaborative Writing in Industry: An Annotated Bibliography for Teachers and Researchers, *The Kentucky English Bulletin, 37*:1, pp. 89–104, 1987.
7. B. Collins and H. Guetzkow, *A Social Psychology of Group Processes for Decision Making,* John Wiley, New York, 1964.

PART I:

Theoretical Overview of the Collaborative Process

Perhaps no other identifiable activity is as often practiced but as little understood as collaboration. Collaboration is little understood, in part at least, because the nature of 20th century industry has demanded that technical communication practitioners develop collaborative schemes and strategies (inevitably "on the run") to manage, improve, and control their own and others' contributions to a company's overall success. Lisa Ede and Andrea Lunsford have noted that 87 percent of respondents from six major professions reported that they at least wrote collaboratively "sometimes" [1]. In fact, since people engaged in collaborative activities frequently do not distinguish them as "collaborative," the number might actually be even higher.

Even though a great number of technical writers and editors engage in collaboration, successful collaboration can be a difficult process. Perhaps we could tentatively define "successful collaboration" as the efficient production of an effective product with a minimum of disruptive conflict. And although some successful procedures have been worked out in the field, far too many practitioners understand that if collaboration is not handled well, the final product can be like that proverbial "horse designed by a committee"—the camel. One of us has, for example, been both participant/observer in a long-term, joint composing effort here on our campus (i.e., a faculty group trying to re-write sections of an operations manual). The difficulties that may result from such factors as participants have varying levels of experience and various agendas, external events disrupting the process, and participants having differing concepts of the task at hand may all contribute to making successful collaboration more difficult to obtain. Many readers, no doubt, have shared similar experiences.

Therefore, a study of the theoretical foundation of collaboration may lead practitioners and academicians to manage more effectively some of these naturally occurring obstacles to collaboration. Undoubtedly, it has been the currency of "social constructionist" thought, best known to writing teachers through the work of Kenneth Bruffee, which has led many academics to begin studying this phenomenon more fully. Clearly this collection has much of its impetus from this academic interest. But in this book and in particular in this Part, we depart from what may be a too frequent occurrence in scholarship— academicians coming along later to study something that practitioners already know how to do. These chapters are ground-breaking in that all four focus to some extent on identifying and investigating current difficulties that surround and stall actual collaborative writing projects. By drawing upon both theory and application, the authors in this section sketch out larger problems and suggest solutions.

David Farkas, in "Collaborative Writing, Software Development, and the Universe of Collaborative Activity," identifies collaborative writing as one of many collaborative activities by demonstrating how collaborative writing is similar to other forms of collaboration in at least six important ways. Further, he explains what makes collaborative writing difficult in comparison to those other activities. And as a particular example of his method, Farkas examines software development in order to glean how that discipline has been successful with its collaborative activity, and identifies strategies which might successfully transfer to collaborative writing.

Timothy Weiss, in "Bruffee, the Bakhtin Circle, and the Concept of Collaboration," notes the ongoing concern between theory and practice and in so doing, "summarizes and critiques [Kenneth] Bruffee's concept of collaborative learning" pointing out what he sees as problematic strands of Bruffee's theory. Drawing upon the ideas of M. M. Bakhtin's Circle and I. Lakatos, Weiss shows how the Circle's ideas regarding "understanding" versus "knowledge" and "object" versus "subject" may clarify Bruffee's theory. Additionally, Weiss draws upon Bakhtin to demonstrate the *potential* value of conflict in the collaborative process. Finally, Weiss concludes with three revisions of Bruffee's underlying theory which may strengthen the validity of Bruffee's observations.

James Weber, in "The Construction of Multi-Authored Texts in One Laboratory Setting," looks at current collaborative theory in light of practice in a large research laboratory. Noting that the role environment plays in the composing process when applied to collaboration has been largely neglected in current scholarship, Weber examines practice and discovers three "areas of knowledge" which cut across both the cognitive and social construction of texts: leadership, goals and content, and methods for resolving disagreements among collaborators. His experience suggests that these are vitally important issues if effective collaboration is to occur. Weber concludes with his key findings and recommendations for more effective practice.

Meg Morgan and Mary Murray in "Insight and Collaborative Writing" draw upon scholarship in such diverse fields as composition, education, philosophy, theology, and non-clinical psychology to offer a look at one key for groups which want to produce better solutions to problems. They argue that collaborative writing groups which are able to arrive at "insight" will, almost necessarily, produce a better solution than if they either fail to achieve an "insight" or somehow avoid doing so. Morgan and Murray discuss the nature of and conditions necessary for producing insight in individuals and show how these conditions may be applied to small problem-solving groups. They find that groups which do not sense some dissonance, do not identify clearly the problem creating that dissonance, and do not confront that problem will likely not succeed. Finally, they offer a "case of non-insight" as an example.

REFERENCE

1. L. Ede and A. Lunsford, Research in Collaborative Writing, *Technical Communication, 32*:4, pp. 69–70, 1985.

CHAPTER 1

Collaborative Writing, Software Development, and the Universe of Collaborative Activity

DAVID K. FARKAS

In recent years, collaborative writing has become a significant area within the discipline of composition studies, and it is now receiving a good deal of attention. We should remember, however, that collaborative writing is not simply a form of writing. It is also one of innumerable forms of human collaborative activity. Just as people collaborate to prepare documents, so do they collaborate to build bridges, climb mountains, govern communities, and perform innumerable other tasks.

Much of the time collaborative activity is carried out transparently and with little or no difficulty. A husband and wife prepare the family breakfast and have no thought that they are engaged in a collaborative task. Other collaborations are complex and difficult affairs in which the need to coordinate the activity of a group of individuals is a prominent consideration. One of the broad theses of this chapter is that collaborative writing is one of these complex and difficult forms of collaboration.

The second broad thesis of this chapter is that collaborative writing has much in common with other forms of collaboration. All forms of collaboration, for example, require participants to develop a common understanding of their goal and methods and to coordinate their efforts toward achieving that goal. But, of course, collaborative writing is much closer to some classes of collaborative activity than to others.

The third broad thesis is that we can learn a great deal about collaborative writing by looking at other forms of collaborative activity, both by observing it directly as it occurs all around us and by adopting and adapting insights and methodologies from disciplines that in various ways address the problem of human collaboration.

My procedure in this chapter is

1. to demonstrate, by means of comparisons with other collaborative activities, why collaborative writing is difficult and to point out that the difficulties in collaborative writing are not unique but rather are shared with broad classes of collaborative activities, and
2. to look closely at one particular collaborative activity—software development—and show how software development can contribute to the understanding and improved practice of collaborative writing and what it has contributed already.

Before proceeding further, I will offer a definition of collaborative writing that consists of four basic forms of writing, along with all possible combinations:

1. two or more people jointly composing the complete text of a document;
2. two or more people contributing components to a document;
3. one or more persons modifying, by editing and/or reviewing, the document of one or more persons; and
4. one person working interactively with one or more persons and drafting a document based on the ideas of the person or persons.

While this chapter applies in at least a general way to all forms of collaborative writing, I most often refer to and I implicitly assume a combination of Forms 2 and 3, a situation in which two or more persons contribute components of a document and one or more persons review and/or edit the document. I de-emphasize Form 4 because it is a somewhat marginal instance of collaboration in that one person is doing all of the actual drafting. I de-emphasize Form 1 because it is less often practiced than the other forms of collaborative writing. Joint composing vastly reduces writers' productivity; the two (or more) writers perform the work of one and, in fact, often produce less material than one writer would. Also, joint composing frequently results in a great deal of disagreement and irritation over individual choices in diction and syntax. This form of collaborative writing is, in fact, primarily useful in situations when the document is short, when agreement on exact wording is crucial, and when there is need to eliminate what would otherwise be an extended round of reviews and edits among the collaborators.

WHY IS COLLABORATIVE WRITING DIFFICULT?

Collaborative writing *is* difficult. I think our collective experience tells us that. Also, the literature supports this idea [1, 2]. Below I offer six reasons why:

1. Highly integrated documents are very complex artifacts.
2. The process of preparing a document becomes more complex when it is performed collaboratively.
3. The writing process generates strong emotional commitments.
4. Documents are reworkable and are subject to infinite revision.
5. Collaborative writers lack fully adequate terms and concepts with which to create a clear and precise common image of the document they wish to produce.
6. It is difficult to predict or measure success.

These six reasons are derived only on an ad hoc basis, not from a comprehensive theory of collaborative writing. None exists. Others studying collaborative writing, therefore, can create a somewhat different list. This list, however, as it is developed below, does articulate much of the difficulty in collaborative writing. In addition, it reveals numerous points of contact between collaborative writing and other collaborative activities.

1. Highly Integrated Documents Are Very Complex Artifacts

Obviously, not all the artifacts produced by collaborative work are complex. A group of children, for instance, will make a crude snow figure consisting of little more than three balls of snow. Any document, however—even a short, routine business letter—is a moderately complex artifact, a web of words that have been chosen on the basis of syntactic, semantic, and pragmatic considerations.

Greater complexity begins to come in longer documents that convey large amounts of information. A requirement for truly great complexity is the high degree of integration—the myriad and complex linkages—among the components of a large document. This integration can take the form of a carefully developed theme or a carefully orchestrated emotional impact that builds and modulates through a long document. It can also take the form of an elaborate system for presenting reference information—for example, a sophisticated reference manual that at each topic in the document directs the reader to all the other places where related information is to be found.

In contrast, there are documents characterized by fairly discrete components and, hence, a "loose" form of collaboration. This category includes a technical report containing sections individually researched and authored by scientists who only coordinated on a few matters. While each section bears upon the overall topic in an appropriate way, the format, organization, depth of treatment, and perhaps even the approach differ from section to section. Such documents may be perfectly satisfactory if no further integration is required or expected by the audience. As collaborative documents, however, they are what we can call "nonproblematic," since the creation of the document entails little more difficulty than would the creation of the parts as separate entities.

There are also documents in which the integration of the components merely follows a set of explicit rules and, hence, is not complex. These are modular documents, such as catalogs and collections of abstracts, whose components are uniform in format, tone, level of detail, etc. When such documents are produced collaboratively, they are also nonproblematical in that the degree of integration is relatively limited. Throughout this chapter I exclude from consideration, unless I indicate otherwise, these nonproblematical collaborative documents.

There are, of course, vast numbers of very complex artifacts, and they are complex for the same reasons that documents are: they contain many elements, and there is a high degree of integration among these elements. An airliner, for example, contains many complex subsystems, and they are highly integrated into complex systems that are closely integrated with one another. A gravel road, in contrast, contains many elements, but the degree of integration among the pieces of gravel is limited.

Any complex system is strongly prone to failure, at least until the flaws are identified and corrected [3]. In the case of a document, failure can take such forms as missing information, illogical or nonfunctional organization, and inconsistencies in point of view or treatment of the subject. Although documents normally do not fail or succeed absolutely, numerous and significant flaws can ultimately result in a nonfunctional document.

2. The Process of Preparing a Document Becomes More Complex When It Is Performed Collaboratively

Many activities that are very simple when performed individually become more complex when done as a group. Digging a hole, for instance, becomes more complex if, for reasons of speed, several people are digging at the hole simultaneously. Preparing a highly integrated document as an individual is unquestionably a complex process. The writer must maintain control of the theme, the organizational structure, the level of detail, the tone, and all other aspects from beginning to end. But the process of preparing a highly integrated whole on a collaborative basis is far more complex.

First, the writing team must develop a common image of what they hope to produce. Then the task must be decomposed among the writers. The work schedule must be carefully coordinated so that individuals are not too often hindered because they are waiting for someone else to finish writing or editing some part of the document. Almost inevitably each writer's "common image" of the document turns out to have been a little different, and almost inevitably each writer drifts still further from that common image as personal habits and predilections assert themselves. As a result, the individual components must be corrected, probably at several stages, for divergences from the common image, as well as for simple errors and deficiencies on the part of the writers.

Given the difficulties described here, we might almost regret that collaborative writing is necessary at all. But it is necessary—so as to bring varied skills and perspectives to a project; to get the work done more quickly; and to effect the sharing of skills and information among team members.

In most instances, complex artifacts must be produced by complex processes, and in general performing these processes collaboratively—while often necessary—adds still more complexity to the process. This greater complexity results in a process that is more prone to failure, failure that can result in a flawed artifact or in the inability to produce any artifact at all. Also, as Frederick Brooks demonstrates, as collaborations grow more complex, the wasted time and effort increases dramatically. For example, Brooks proves that the extra orientation time and coordination effort required when more people are added to a complex collaborative project will negate the savings in time that the project leaders were hoping to achieve [4, pp. 13-26]. Complex collaborative processes, however, are an inextricable part of our world, and so we must learn to mitigate the problems they pose and perform them as successfully as possible.

3. The Writing Process Generates Strong Emotional Commitments

All work has an affective component; people will be satisfied, rebellious, or uninterested under a given set of circumstances. Certain kinds of work, however, generate strong personal commitments, the feeling that something is "my handiwork," "my baby." Strong emotional commitments are especially likely in the case of work, such as design work, in which the individual can make a distinctive and identifiable contribution to the final product.

Writing seems unusual in the strength of the commitments it generates. Even in the case of routine documents, writers often take pride in their writing or harbor sensitivities about its deficiencies. They tend to hold strongly to the precepts of style and usage they learned in school or picked up on the job, and they are not relaxed about the opinions that others may have about the precepts they follow or their overall writing ability.

This strong commitment severely complicates collaborative writing in many instances. People often staunchly resist the changes made or asked for by editors and reviewers. They hold fast to their own image of the finished product and want the other components to conform to that image. They regard their component of a collaborative work as a private document and assume the prerogatives of traditional authorship. They sub-optimize their component at the expense of the whole.

Ideally, the strong commitment generated by writing can be properly channeled into teamwork and pride in the collaborative product. This is perhaps the major challenge to the management of collaborative writing.

4. Documents Are Reworkable and Are Subject to Infinite Revision

Many collaborative activities cannot be repeated once they've been performed. A football team, for instance, gets no second chance at the play that cost them the championship. The cockpit crew of an airliner cannot recapture the moment in which a fatal error was committed.

A much larger number of collaborative tasks do enjoy either complete or at least partial reworkability. In the production of goods, we often enjoy reworkability; if a component is set in the wrong position, the assembler can simply reposition it. In film making, there is partial reworkability: there is no need to redo an entire film if one scene is shot badly, but the scene itself must be redone. Intellectual products such as plans, designs, decisions, and documents are fully reworkable. No materials have been cut, no chemical processes have been initiated; nothing need be declared complete until everyone is satisfied with it.

Certainly reworkability is a boon. No one wants to engage in actions they cannot recall. But reworkability has a downside for collaborative writing as well as for many other activities. In the professional world it is common for bosses and various reviewers (a group of after-the-fact and often unwelcome collaborators) to ask for a great many modifications in a draft document. For instance, proposals prepared by engineering teams at a major Northwest manufacturing corporation may undergo as many as 50 review and editing cycles. While the review process is often necessary as a means of directing the work of subordinates, review changes also come about simply from the universal impulse people have to put their own stamp on a project [2].

Excessive reworking has various negative consequences:

a. Valuable time is lost and the project falls behind schedule. Sometimes, in fact, it is only a looming deadline that finally curtails the review process.

b. The instability that is introduced into the developing document severely taxes the writing ability of team members. A writer has difficulty composing when he/she knows (or suspects) that the other components are being or will be significantly changed.

c. Writers become frustrated and demoralized when their work is endlessly changed. They grow resentful at what they regard as criticism of their work, and they disengage themselves from their writing, as in "What does it matter what I write, they'll only change it anyway."

d. The document's conceptual integrity is often lost. High-level reviewers may introduce (or ask for) changes in a large document without having gotten to know it intimately and without fully appreciating that changes in one place may cause problems elsewhere in the document [5,6]. Ultimately severe incompatibilities are introduced into the document, and if enough reworking has taken place in a short period of time, no one may retain a clear grasp of the current state of the document.

In many collaborative writing situations, the forces that militate toward excessive reworking are often very strong, but controlling this tendency is crucial for the success of the project [7].

5. Collaborative Writers Lack Fully Adequate Terms and Concepts with Which to Create a Clear and Precise Common Image of the Document They Wish to Produce

A civil engineer can specify exactly how he/she wants a bridge built, and all the work groups involved in the project can derive an almost identical image of what is to be constructed. The members of an auditing team typically understand all aspects of the numbers and accounting rules they are dealing with, and so they can pass information easily to one another.

But writers work with one of the most elusive and poorly understood materials—language—and so they generally use concepts and terms that are imprecise and deficient for communication among team members. When a proposal manager attempts to convey to a team of researchers the tone she is seeking, she may say that she wants the proposal to be "punchy" and "hard-hitting," but also "casual" and "friendly." These phrases, however, are only metaphors and they do not provide much guidance to the individual writers. Certainly, the use of such phrases is not enough to ensure that each member will produce a component similar to the others in tone.

Part of this problem is the lag with which academic research moves into the world of work. We may soon hear proposal managers and other writing team leaders specifying the product they want through references to propositional units, denominalizations, embedded clauses, and anaphoric references in much the same way that they often rely upon Gunning and Flesch. But the larger part of the problem, I believe, is that the terms and concepts that would really enable a manager to provide a clear and precise image of an as-yet-unwritten document simply do not exist. Academic rhetoricians themselves still talk largely in metaphors when they plan and execute collaborative writing projects. Beyond a certain level, it may be impossible to precisely specify a common image of a collaborative document; writers will always bring unique life experiences and personal lexicons to any writing project they embark upon. But we can go a long way toward developing terms and concepts that could be put to practical use in specifying such matters as tone, style, detail of treatment, and method of integrating the parts into the whole.

The situation faced by writing teams trying to establish a common image for an emerging document is not so different from that faced by any group of design professionals whose product has an important aesthetic dimension—say, an architect specializing in exteriors, one specializing in interiors, an interior decorator, and a landscape architect—working together on a corporate headquarters building. They too will often be using highly metaphorical design terms, such as harmony, energy, rhythm, and balance, when trying to achieve a high level of integration in the total design.

Both writing teams and design teams compensate to some degree for limitations in their design vocabularies by using preliminary and partial representations of the finished product. A proposal manager may produce some sample pages (in addition to an outline) for the benefit of the team members. The design team members can create thumbnail sketches, computer graphics, and balsa models. These representations can be debated and modified and when consensus is finally reached can serve as a common image of the finished product. Because visual images can be grasped more quickly than text and because they can generally be modified more easily than text can be re-written, the representations of architects probably serve more adequately than those of writers.

6. It Is Difficult to Predict or Measure Success

When a team of mechanical engineers tests the prototype of the automatic packaging machine they are designing, the device will fail if there is a major flaw in their work. Also, there are observable signs if the vibration is too great, if parts are out of balance, or if there are other problems in the performance of the device. Thus the engineers can immediately set about to correct the deficiencies in their design. This readily obtained, objective feedback greatly increases the likelihood that the final product will be successful. The ready availability of useful feedback is characteristic of a great many activities, including one linguistic activity: a group of people can collaborate productively on a crossword puzzle because the crossword puzzle provides a means for the collaborators to easily test and reject incorrect suggestions.

In contrast, a group of marketers cannot predict reliably whether their new brand of toothpaste or perfume will succeed with consumers, and a group of economists cannot be sure that the public will change its buying or saving habits in accordance with a new economic plan. Similarly, documents take their meaning and achieve their effects in the minds of readers, and therefore it is difficult to predict whether a document will achieve the intended results. Often, in fact, it is difficult to determine whether a published document *has* achieved the intended results. Usability testing along with other forms of evaluation is, of course, a valuable method for achieving objective and presumably reliable feedback. But testing is expensive, often yields data of limited usefulness, and requires time that may not be available [8,9]. Furthermore, most forms of document testing cannot be performed until much of the document has been created.

It is often difficult to resolve differences among collaborators when the question at issue will not be settled through objective testing. Also, the opinion that prevails may not be the best. If, as I noted in the section on reworkability, incompatibilities are introduced by highly placed reviewers, they are likely to destroy the conceptual integrity of the published document.

Although I have been focusing on why collaborative writing is difficult, I have also pointed out that other collaborative activities share in these same difficulties. The six reasons I give for the difficulty of collaborative writing are also, then, reasons—though certainly not the full set of reasons—why many collaborative activities are difficult. Many collaborative activities result in highly integrated and complex artifacts, are produced through very complex processes, and are infinitely reworkable. In a smaller but still significant number of collaborative activities, collaborators form strong emotional commitments to their contributions to the finished product, lack precise terms and concepts, and face difficulties in predicting and even measuring success. Aesthetic design work is one collaborative activity that shares in each of the six difficulties cited.

Despite these difficulties, the fact that collaborative writing has much in common with other forms of collaborative activity is good news. It is a strong sign that we can address the difficulties posed by collaborative writing by looking at other forms of collaborative activity. I am not speaking now of looking at the various settings in which collaborative writing is performed—although this is certainly an important research direction [10, pp. 23-53]. Rather, I am speaking of the value of studying a range of collaborative activities, activities that may not entail any writing at all. Among the many activities that can be profitably examined are sports, business, ensemble music, and software development.

We can examine the behavior of sports teams, perhaps through our own participation, for insights into team behavior and for methods of achieving team unity and peak performance when the occasion demands. Also, much relevant material can be found in the literature of athletics, sports, and coaching. For example, a group of professionals beginning a challenging and arduous writing project might adapt the methods developed by coaches for use in pre-competition meetings [11, pp. 117-137].

The world of business is and certainly should be studied as one of the main settings in which collaborative writing takes place [2, 12]. But apart from this, business is a highly relevant collaborative activity in which complex projects are planned and executed, structures for group work are established, and the efforts of individuals are coordinated. In addition, the supporting discipline of management is a fertile source of theory and methods that can be applied to collaborative writing. By examining the world of business and the discipline of management, we may, for example, find better ways of scheduling and coordinating the writing of documents as well as better ways of organizing the review process.

Ensemble music is a collaborative activity in which the medium, like language, is subtle and elusive. Also, apart from the technical aspects of music, the terms and concepts musicians work with are highly metaphorical and lacking in precision. Most important, each musician, while engaging in a very personal and expressive form of communication, must blend his or her individual voice into a group product. In the case of orchestral music, the score largely constrains the choices of the individual musicians, but a conductor is still employed to coordinate aspects of the musicianship. In the case of improvisational music, such as jazz, each performer proceeds with considerable autonomy but must maintain ongoing communication with the others. The very frequent success enjoyed by ensemble musicians of all kinds indicates that there is much in their methods and traditions to be carefully examined by those with an interest in improving collaborative writing.

Another collaborative activity that should be studied is software development along with its associated discipline, software engineering. In this instance, however, I will do more than glance at its relevance to collaborative writing. Rather, the nature of the contribution that software development can make, and has already made, is the subject of the second part of this chapter.

SOFTWARE DEVELOPMENT

Software Development and Collaborative Writing

Although computer programs are produced by individual programmers, most commercial software is sufficiently large to require a team of programmers. Software development, therefore, is by and large a collaborative activity.

As a collaborative activity software development has significant similarities to collaborative writing. First of all, programmers, like writers, create long and complex strings of language—computer code. The language of programmers, of course, is a formal rather than a natural one, and its primary "audience" is a group of electronic components. But the electronic components, like human beings, make demands upon the language they deal with, and because the components require logic and coherence, each programmer's part must fit properly into the whole. Furthermore, computer code must be readable not only by electronic components but by human beings as well. Otherwise, it is impossible for members of programming teams to work with each other's code or for finished programs to be maintained by other programmers at a later date. Programmers, in fact, speak about good and bad programming "style" in somewhat the same way that writers do [13, 14].

Software development also faces many of the same difficulties that collaborative writing does. In fact, if we adapt the list of six difficulties in collaborative writing to software development, we get a meaningful picture of the similarities and differences between the two.

1. Large computer programs are very complex artifacts and are highly prone, at least initially, to failure.
2. Because large computer programs are produced collaboratively, the process itself is extremely complex and fraught with difficulty. The work, first of all, must be decomposed among the programmers, and action must be taken continually to correct for divergences from the common image. In contrast to writing, software project leaders can specify this image precisely, but, as Edward Yourdon notes, wherever the specification is incomplete, individual programmers will tend to make individual design decisions [15, p. 150]. Furthermore, as Yourdon notes again, "communication problems between programmers become unmanageable on large projects" [15, p. 150].
3. Programmers, like writers, form strong emotional commitments to their work. They become committed to their own programming techniques and the code they produce. Furthermore, they tend to regard their segments of code as "private masterpieces" rather than as part of a group product [15, p. 172].
4. Like writing, code is infinitely reworkable and like writing is subject to successive reworking. Managers and customers change specifications, and developers must struggle to keep the project on schedule and within budget and to maintain conceptual integrity [16, pp. 151-240].
5. In clear contrast to collaborative writing, software development enjoys precise terms and concepts. This precision extends in many cases to the language of mathematics. These terms and concepts enable software developers to clearly and precisely specify what they wish to create.
6. In contrast to the situation faced by writers, software developers receive feedback that is objective. This feedback, however, is only somewhat more readily obtainable than the feedback writers seek through usability testing. If there are major flaws in the coding, if—let us say, serious incompatibilities have been introduced at the last minute—these flaws become readily apparent, because the program will crash. On the other hand, subtler problems conceal themselves deep within the code; and if not discovered and corrected through a process of rigorous software testing, they will reach the user in the form of a "buggy" product.

The Discipline of Software Engineering

The difficulty in creating large computer programs became increasingly apparent in the 1960s, and led to the emergence of a new discipline devoted to the support of software development. This discipline is software engineering. Utilizing engineering methodology, management theory, and findings from psychology, software engineers attempt to improve the efficiency with which computer programs are designed, written, tested, and maintained. Because inefficiencies in large software development projects are so expensive and so visible, software engineering is an active, well-funded discipline.

Because software development is a collaborative activity, software engineering is in large part the study of human collaboration in a particular setting. So, for example, Frederick Brooks' *Mythical Man-Month* is both a pioneering work in the field of software engineering and a classic study of human collaboration [4]. As we will see, collaborative writing has already benefited significantly from some of the work done in software engineering, and, can benefit still more. The aspects of software development and software engineering most relevant are 1) models for software development, 2) tools for collaboration, 3) metrics, and 4) visibility and access to resources.

Models for Software Development

Programming teams employ conceptual models to develop software. These models have close analogs to collaborative writing. The classic model is "top-down design." The first steps are to ascertain system requirements, perform top-level design, and then perform successively more detailed design. When this planning process is complete, programmers write their components of the program. The completed components are then integrated into a whole, and the program is tested in order to determine if further work is necessary. One criticism of this model is that it remains a set of abstract plans for too long. Consequently, many developers modify the model by calling for "draft" versions of each component midway through the development process. These components are tested and evaluated, so that more "proven" versions of each component are available for integration into the whole [17, pp. 22-25].

"Rapid prototyping" is a model that requires less detailed preliminary planning. Instead it calls for the rapid and inexpensive development of a partly functional prototype that has the "look and feel" of the finished product. This prototype is evaluated by users and then refined by the programmers in successive cycles until the program is complete. The advantage of this more casual approach to software development is that users get to critique early versions of the program before the design has solidified [18, pp. 26-27].

Another model is "structured design" [15, pp. 86-100]. Its underlying concept is "modularity." Modularity, however, is a broad concept that pervades many models of software development. The idea behind modularity is to reduce the overall complexity of a program by breaking it down into a set of semi-discrete units whose relationships with one another are fairly easy to define. This contrasts to "spaghetti code," in which any segment of code can have connections to any other segment.

Other models focus upon the collaborative arrangements of the development team. Among these is the "chief programmer team" model. Here one very superior programmer does all the critical programming on a project, but has a very complete support staff—assistant programmers, an administrator, software testers, and so forth—that enables her to get a large job done in a reasonable length of time. The idea is to reduce the complexity of the software development process by drastically reducing the amount of communication and coordination that is necessary among team members and to ensure that at least one person has a complete grasp of the project. One major problem with the model, however, is that only relatively small projects can be undertaken with this method [4, pp. 29-37] .

A very different model focusing on collaborative arrangements is that of the ego-less programming team. Here the goal is to achieve synergism and to eliminate the problems caused by emotional commitments, including the tendency to view one's work as a "private masterpiece." In an ego-less programming team, for example, team members have free access to each other's code. In addition, objective discussion of each person's work is formalized through the use of special meetings known as "walkthroughs." The group rather than individuals assumes responsibility for the success of the project, and the emphasis is on the achievement of the team rather than individual contributions [15, pp. 171-172] .

All of these models have analogs in writing. Top-down design is akin to having a writing team develop and agree upon the top-level entries of an outline and then proceed to lower-level entries. It might include drafting preliminary versions of chapter introductions and other high-level components. One conception of how to apply rapid prototyping to the domain of writing is for the writers to produce a quick rough draft or even a dictaphone recording of representative sections of the document and solicit comments from managers or trial readers before refining the design.

Modular documents do exist—catalogs and collections of abstracts are examples. Indeed, although modularization can only be used in certain communication situations, the approach simplifies the collaborative process vastly, so much so that early on in this chapter modular documents were classified as a nonproblematic form of collaborative writing.

There have no doubt been innumerable instances in which collaborative writing has been carried out very approximately along both the model of the chief programmer team and the ego-less programming team—certainly there are times when it is desirable and feasible to simplify collaborative writing using the marginal Form 4 described at the beginning of this chapter, and certainly it is often desirable to attempt to inculcate team spirit among the group of writers and to appropriately channel the strong emotional commitment generated by the writing process. But here is the key point: because of the work of software developers and software engineers, these software development models have been carefully formulated and debated, systematically applied, and empirically studied. There is a highly analytical literature that describes these models and their underlying concepts and explains and debates their merits and deficiencies. On the other hand, the analogs in collaborative writing are far less carefully developed and less often studied. They exist primarily in the collective experience of individual writers and writing team leaders. The literature is scant and very often anecdotal.

Collaborative writing may require its own distinct development models, but there is certainly much to be borrowed from the models of software development and the analytical literature that accompanies them. Of note in this regard are Ronald Guillemette's rationale for adapting rapid prototyping to documentation writing [19] and Edmund Weiss' plan, which incorporates a variety of software engineering concepts, for preparing computer documentation using a modular format that both satisfies the information needs of readers and simplifies the collaborative writing process [20].

Computer Tools for Collaboration

Because collaborative writing and software development are very complex processes, both writing teams and programming teams face the challenge of keeping track of what has been produced, ensuring that changes are made only by those with authorization, and making the most current version of the product (as well as earlier versions) available to all who need to view it. Software developers are tool-builders, and since the 1960s they have been creating and refining a whole class of tools for this purpose [4, pp. 132-133]. These "configuration management" tools are the computer equivalent of a project librarian. If, for example, a programmer wants to see the most recent version of a segment of the project code, the computer delivers it, indicates by whom and when the last revisions were made, and can even display just those lines in which changes were made.

These tools have now migrated into the world of technical publications, and the most sophisticated electronic publishing systems now possess elaborate configuration management capabilities [21, 22, 23]. These capabilities are beginning to appear on microcomputers, both as advanced features of high-end word processors or as separate products. In this form, they will soon be available to almost everyone who engages in collaborative writing.

Software developers, especially those working in research environments, have also devised a broader range of tools to support their activities and to advance our understanding of collaborative work. There are very interesting noncommercial systems which facilitate face-to-face meetings [24, 25] and which facilitate meetings among people working at different locations [26, 27]. There are also systems designed to facilitate collaborative writing in new ways. Neptune, for example, is an experimental hypermedia system that enables the members of writing teams to share text across a network and, more significantly, to manipulate and view this pooled material in ways that help them visualize the emerging document [28]. The practice of collaborative writing will certainly improve as these tools are refined and commercialized. In the meantime, the computer science literature contains articles describing these tools, the theoretical assumptions they embody, and the findings of experiments that involve their use. This literature provides valuable insights concerning both collaborative writing and collaborative work in general.

Development of Metrics

Software engineers have developed metrics, appropriate measurements, that they use to better understand and plan their work. A very crude but useful metric is the size of the project measured in lines of code. It is helpful, for example, to know that the program you are about to produce is estimated at 20,000 lines of COBOL code. Another crude metric is programmer productivity measured in lines of code per unit time. The inadequacy of these metrics, however, is easy to see: one 20,000 line program may be much more complex to write than another, and one programmer may write 300 lines of easy code in a day whereas another may produce 50 lines of very difficult code. Software engineers, therefore, have developed very sophisticated, often highly mathematical, metrics that give them a much more precise understanding of the nature of a software development project [29, 30].

Writing teams understand and use both the metric of document length and the productivity metric of pages produced per unit time. But we need to develop more sophisticated metrics in order to better plan and schedule writing projects and to distribute the workload more evenly and with more attention to the special abilities of individual writers.

We would benefit from predictive metrics that measure the writer's familiarity with the source material or the amount of time the writer will spend getting information from technical experts. Other predictive metrics might measure structural characteristics of an unwritten document, based on an outline or an understanding of its organization. For example, a metric that indicated the degree of interrelatedness of the parts of a document would help predict how quickly the writing would go, what the likelihood was of leaving out necessary information, how the document should be decomposed among a team of writers, and which components should go to the most capable members of the team.

Software engineering has something to contribute to the development of sophisticated metrics for writing. To begin with, we can look to software engineering for hints as to what sorts of metrics are desirable and how to develop them. The applicability of software metrics is, of course, limited by the difference between formal and natural language and the fact that the primary audience for code is not a human being. Consequently, metrics for writing will be based largely on concepts of language and discourse that will be drawn from linguistics and rhetoric and an understanding of composing processes and reading processes that will come from psychology. But software engineering metrics may provide the mathematical relationships that turn distinctions about language and an understanding of psychological processes into useful quantitative measurements.

Visibility and Access to Resources

Although professionals in many fields do some incidental programming, almost all large software development projects are created by full-time programmers. Software development, therefore, has become a distinct profession.

Furthermore, the profession has been successful in achieving visibility and gaining access to resources. Corporations and government agencies fund research in software development, and indeed the discipline of software engineering has emerged to support this activity. Corporations and other producers of software have recognized the importance of programmer productivity and quality software, and hire consultants and seminar instructors to help them create an appropriate environment for software development. Books such as Paul Licker's *The Art of Managing Software Development People* serve the same function, explaining to managers that programmers have specific needs and require specialized management structures and procedures [31].

On the other hand, although collaborative writing is an all pervasive activity practiced throughout the professional world, it is largely invisible. This is because it is primarily practiced as an ancillary activity by bankers, marketers, chemists, social workers, and so forth. In many instances, the wasted time and energy, the animosities that have arisen, and even the poor quality of the finished document are forgotten shortly after the project has ended. The full-time collaborative writing professionals, primarily technical communicators and other corporate publications professionals, make up only a small contingent. Furthermore, in the nation's English departments, the primary home for the study of writing, the concern with collaborative writing is only recent.

The main theme of this chapter is that collaborative writing is difficult, that its difficulties are largely shared with other activities, and that we can look at these activities and profit. Looking at software engineering, one thing we learn is the need to teach about collaborative writing as well as study it. We need to ensure that writing teams have adequate visibility and access to resources.

Corporations should come to realize that writing, especially in groups, is a complex and delicate process and that special management procedures are necessary. For example, the document review process should be conducted with sensitivity to the burden it places on writers, and special procedures, such as some variation on the ego-less programming team model should be adopted. Corporations should also recognize the need for special resources, such as specialized computer equipment and the time and money necessary for usability testing. Finally corporations and government agencies should realize the value of well-funded research in the area of collaborative writing.

Academics and practitioners working together should be able to achieve a noticeable effect in the corporate world. But whatever our degree of success over the near term, the recent trend toward emphasizing collaborative writing in the schools is very promising. If this trend continues and strengthens, we can expect future generations of professionals to understand more about the practice of collaborative writing and to give it greater support within their organizations. This is important, for it will help create a more productive and congenial workplace.

REFERENCES

1. L. S. Ede and A. A. Lunsford, Why Write . . . Together: A Research Update, *Rhetoric Review, 5*:1, pp. 71–77, 1986.
2. J. Paradis, D. Dobrin, R. Miller, Writing at Exxon ITD: Notes on the Writing Environment of an R&D Organization, in *Writing in Nonacademic Settings,* L. Odell and D. Goswami (eds.), Guilford Press, New York, 1985.
3. J. Gall, *Systemantics: How Systems Work and Especially How They Fail,* Pocket Books, New York, 1976.
4. F. B. Brooks, Jr., *The Mythical Man-Month: Essays on Software Engineering,* Addison-Wesley, Reading, Massachusetts, 1975.
5. D. K. Farkas and N. J. Farkas, Manuscript Surprises: A Problem in Copy Editing, *Technical Communication, 28*:2, pp. 16–18, 1981.
6. D. K. Farkas, The Concept of Consistency in Writing and Editing, *Journal of Technical Writing and Communication, 15*:4, pp. 353–364, 1985.
7. E. Gold, Don't Let the Approval Process Spoil the Book, *Simply Stated, 49,* pp. 1–2, September 1984.
8. G. M. Schumacher and R. Waller, Testing Design Alternatives: A Comparison of Procedures, in *Designing Usable Texts,* T. M. Duffy and R. Waller (eds.), Academic Press, New York, 1985.
9. P. Wright, Is Evaluation a Myth? Assessing Text Assessment Procedures, in *The Technology of Text,* Volume 2, D. H. Jonassen (ed.), Educational Technology Publications, Englewood Cliffs, New Jersey, 1985.
10. J. M. Lauer and J. W. Asher, *Composition Research: Empirical Designs,* Oxford University Press, New York, 1988.
11. J. Syer and C. Connolly, *Sporting Mind Sporting Body: An Athlete's Guide to Mental Training,* Prentice-Hall, Englewood Cliffs, New Jersey, 1987.
12. S. Doheny-Farina, Writing in an Emerging Organization: An Ethnographic Study, *Written Communication, 3*:2, pp. 158–185, 1986.
13. B. Shneiderman, *Software Psychology: Human Factors in Computer and Information Systems,* Winthrop Publishers, Cambridge, Massachusetts, 1980.
14. B. W. Kernighan and P. J. Plauger, *The Elements of Programming Style,* McGraw-Hill, New York, 1978.
15. E. Yourdon, *Managing Structured Techniques: Strategies for Software Development in the 1990's,* 3rd Edition, Yourdon Press, New York, 1978.
16. W. L. Bryan and S. G. Siegel, *Software Product Assurance: Techniques for Reducing Software Risk,* Elsevier, New York, 1988.
17. M. W. Evans and J. Marciniak, *Software Quality Assurance and Management,* John Wiley and Sons, New York, 1987.
18. L. S. Levy, *Taming the Tiger: Software Engineering and Software Economics,* Springer-Verlag, New York, 1987.
19. R. A. Guillemette, Prototyping: An Alternative Method for Developing Documentation, *Technical Communication, 34*:3, pp. 135–141, 1987.
20. E. H. Weiss, *How to Write a Usable User Manual,* ISI Press, Philadelphia, 1985.
21. R. A. Grice, Using an Online Workbook to Produce Documentation, *Technical Communication, 30*:4, pp. 27–29, 1983.

22. K. Nichols and L. Duggan, Sharing Common Source Files for Documents: The Agony and the Ecstasy, *Proceedings* of the 35th International Technical Communication Conference, Philadelphia, May 1988, Society for Technical Communication, Washington, D.C., pp. ATA 137–140, 1988.

23. *Introducing Change Control*™, Context Corporation, Beaverton, Oregon, 1987.

24. M. Stefik, G. Foster, D. G. Bobrow, K. Kahn, S. Lanning, and L. Suchman, Beyond the Chalkboard: Computer Support for Collaboration and Problem Solving in Meetings, *Communications of the ACM, 30*:1, pp. 32–47, 1987.

25. M. Begemen, P. Cook, C. Ellis, M. Graf, G. Rein, and T. Smith, Project NICK: Meetings Augmentation and Analysis, *Proceedings* of the Conference on Computer-Supported Collaborative Work, Austin, Texas, December 1986, MCC Software Technology Program and the ACM, pp. 1–6, 1987.

26. S. R. Ahura, J. R. Ensor, and D. N. Horn, The Rapport Multimedia Conferencing System, *Proceedings* of the Conference on Office Information Systems, Palo Alto, California, March 1988, ACM and IEEE, pp. 1–8, 1988.

27. K. Lantz, An Experiment in Multimedia Conferencing, *Proceedings* of the Conference on Computer-Supported Collaborative Work, Austin, Texas, December 1986, MCC Software Technology Program and the ACM, pp. 267–275, 1987.

28. N. M. Delisle and M. D. Schwartz, Collaborative Writing with Hypertext, *IEEE Transactions on Professional Communication, 32*:3, pp. 183–188, 1989.

29. S. D. Conte, H. E. Dunsmore, and V. Y. Shen, *Software Engineering Metrics and Models,* Benjamin/Cummings, Menlo Park, California, 1986.

30. B. W. Boehm, *Software Engineering Economics,* Prentice-Hall, Englewood Cliffs, New Jersey, 1981.

31. P. S. Licker, *The Art of Managing Software Development People,* John Wiley and Sons, New York, 1987.

CHAPTER 2

Bruffee, the Bakhtin Circle, and the Concept of Collaboration

TIMOTHY WEISS

"Word is a two-sided act."
V. N. Voloshinov (1929)

"Writing is not a private act. It is an aspect of social adaptation."
Kenneth Bruffee (1985)

In any communication, speaker and listener, or writer and reader, in conjunction with a particular, shaping context, collaborate in the making of meaning. Their exchange is a "two-sided act" through which the listener, whose active understanding holds "the embryo of an answer," becomes the speaker, and vice versa [1, p. 22; 2, p. 68; 3, p. 86].

The inherently collaborative nature of communication is the topic that this chapter will amplify by bringing together the concept of collaborative learning and writing of composition specialist Kenneth Bruffee and the theories of language, sign, and communication of Soviet linguist, literary theorist, and philosopher M. M. Bakhtin and his collaborators [4], V. N. Voloshinov and P. N. Medvedev. Bruffee is familiar to teachers of technical and professional communication through his articles on collaborative learning–writing in *College English* [5]. Bakhtin and the intellectuals who gathered around and collaborated with him, a group referred to as the Bakhtin Circle, is much less familiar to teachers and theorists in our field. But during the past decade Bakhtin has acquired a reputation as one of the major thinkers of the twentieth century; although his writings fall more within the domain of linguistics, semiotics, and literary theory, he has much to say on the broad subject of communication and merits reading and scholarly attention by theorists and researchers in our field. He has much to say, for instance, about audience, the relationship of scientific and technical writing to other genres of writing, and tone. This chapter introduces teachers and theorists of technical and professional communication to the insightful, heuristic ideas about language and communication of the Bakhtin Circle; it brings these complex, sometimes labyrinthine, ideas to bear on the theory underlying Bruffee's concept of collaborative learning-writing, in an attempt to shed light on dark areas of that theory and to suggest revisions of problematic aspects of it. Bruffee's theoretical rationale for and concept of collaborative learning-writing will thus provide a context for a discussion of some of Bakhtin's principal ideas on language and communication; through this interaction between the familiar and the new, our understanding of the theoretical framework of collaborative learning-writing will be deepened.

This deepening of understanding, Bakhtin would say, constitutes the ultimate reason for thinking about or reflecting on anything; but practically, pedagogically speaking, why should teachers of technical and professional communication invest their efforts in theory, specifically a theory of collaboration? At least three practical reasons exist for thinking about the theoretical framework of collaboration. First, collaborative learning-writing is still a "newfangled" teaching approach, and the teacher who tries it may not understand it and thus may become "tangled" by it. Second, collaborative learning-writing has been presented as a teaching approach for lower-level, composition-oriented courses, not for upper-level technical and professional communication courses in which many students will have a background in a scientific or technical field; these students will question the theory of knowledge underlying the concept of collaboration and may be unwilling to engage in the same kinds of activities that collaborative learning-writing recommends for freshman English students. And third, the theory of knowledge upon which Bruffee bases collaborative learning-writing is problematic in a double sense: it is oriented toward agreement and accord at the risk of devaluing difference and disagreement; it is humanities-oriented rather than sciences-oriented, and this, again, may translate into problems in the technical and professional communication classroom. Teachers need to think about the theoretical foundations of collaborative learning-writing because, among other reasons, it is not good enough to tell our students that they should collaborate on assignments simply because professionals in business, government, science, and technology collaborate on projects large and small. The classroom and the work place are different worlds with different expectations and different rewards.

Having noted the vital link between theory and practice, I will go forward to discuss theory and theory alone in this two-part chapter. Part One summarizes and critiques Bruffee's concept of collaborative learning-writing; it points out what are weaknesses of this concept and of the theory underlying it. Part Two introduces readers to the ideas about language and communication of the Bakhtin Circle and views Bruffee's concept of collaborative learning-writing from the perspective of the Circle's writings. The chapter concludes with a reiteration of key points as they apply to a deeper understanding of a concept of collaboration.

A SUMMARY AND CRITIQUE OF BRUFFEE'S CONCEPT OF COLLABORATIVE LEARNING-WRITING

In *A Short Course in Writing: Practical Rhetoric for Teaching Composition through Collaborative Learning* (1985) Bruffee defines collaborative learning as an approach to teaching that uses group work, attempting to develop and focus the "resource" of "peer influence" [6, p. 1]. Unlike other approaches, the collaborative method assumes that reading and writing are "social" rather than "solitary, individual acts"; its goal is "to help students learn to write better" through engaging their membership in "an active, constructive community of writers and readers" [6, p. 1].

The origins of the concept of collaborative learning-writing are threefold and lie in the experimental teaching and writings of M. L. J. Abercrombie and Edwin Mason, author of *Collaborative Learning,* 1970, in Bruffee's experience as a composition teacher, and in the growing interest in the United States during the 1960s and 70s in alternative methods of teaching [6, p. 337].[1] Abercrombie, a British biologist, used collaboration to teach diagnostics to medical students because she discovered that students working in groups learned "faster and more reliably" than students working individually [6, p. 5]. Bruffee relates Abercrombie's experiment to American composition teachers' experiments in the 1960s and 70s, concluding that: "the best way to learn good judgment is to practice making judgments in collaboration with other people who are at about the same stage of development ..." [6, p. 5]. Bruffee's own teaching experiences, plus empirical studies on group decision making, convinced him that group work helps students learn to make judgments faster and to retain this ability when alone, outside of the group [6, p. 5].

Bruffee puts forward the rationale and theoretical framework for collaborative learning-writing in two essays, "The Structure of Knowledge and the Future of Liberal Education" (1981) and "Liberal Education and the Social Justification of Belief" (1982). In these essays Bruffee calls for a new approach to teaching in the humanities to match "the revolution in [the] conception of knowledge" that twentieth-century physics, mathematics, and philosophy have brought about [7, p. 178]. Bruffee declares that liberal education, in order to be

---

[1] Bruffee discusses the origins of collaborative learning in detail in *A Short Course* and in "Collaborative Learning and the Conversation of Mankind" [5, 6].

revitalized, must "institutionalize the results" of this revolution [7, p. 178]. The rationale for a new approach to teaching is based on the shift from a conception of "determinate" to "indeterminate" knowledge and on "the social implications" of our methods of learning.

Bruffee's concept of collaborative learning-writing is grounded in an epistemology, or theory of knowledge, which differentiates "determinate" and "indeterminate"; that is, that it is *not* "hierarchical," externally determined, and individually attained [7, p. 178]. He cites Einstein's theory of relativity, upon by groups or communities [7, pp. 180-181]. Bruffee states that twentieth-century physics and mathematics have demonstrated that all knowledge is "indeterminate"; that is, that it is *not* "hierarchical," externally determined, and individually attained [7, p. 178]. He cites Einstein's theory of relativity, Heisenberg's uncertainty principle, and Gödel's Theorem as intellectual break-throughs which have altered our concept of knowledge: Einstein "cast doubt on our ability to measure things," Heisenberg "on our ability . . . to observe them," and Gödel on "the symbolic system" within which observations and measurements are expressed. In terms of epistemology, "mathematics, like God, is dead," Bruffee proclaims [7, p. 180].

Referring to Thomas Kuhn's *The Structure of Scientific Revolutions* (1970) and Richard Rorty's *Philosophy and the Mirror of Nature* (1979), Bruffee defines knowledge as a process, a social process, not a product. Echoing Kuhn he writes that scientific change, or progress, is not determined by "a process of gaining new information," but rather by "a social process of shifting communities of scientists" [6, p. 338]. According to Bruffee, scientific knowledge "is what is accepted by an assenting community of scientists," who establish natural laws "by argument and agreement, just as the notoriously vulnerable principles and theories of humanistic studies and social sciences." These natural laws are "consignable ignominiously to the intellectual dustbin whenever the consensus of the assenting community is broken" [7, p. 181]. Echoing Rorty, Bruffee argues that knowledge is "socially justified belief" [6, p. 338]. All disciplines, even "hard-core" natural sciences, all specific discourse communities, determine what is and is not knowledge by consensus. Bruffee takes this epistemological position because, as he sees things, unless knowledge can be equated with belief, an activity like collaborative learning can be nothing more than a collaborative opinion-making, a proverbial "blind leading the blind" [6, p. 338]. This is indeed the heart of the matter, which we will turn to presently in our critique.

Bruffee concludes that a revitalized liberal education would focus on the processes by which individuals in communities reach consensus, on "decision-making, evaluation, analysis, synthesis, establishing or recognizing conceptual frames of reference and defining 'fact' within them" [7, p. 181]. Bruffee defines collaborative learning-writing as the process of negotiating and renegotiating "the frames of reference which fence in and make meaningful what, for the time being, we agree to call knowledge" [7, p. 183]. Since language is both the tool and the stuff of collaboration, it is in and through language that communities establish knowledge "by negotiating consensus and assent" [7, p. 185-186]. Thus knowledge and "truth"—Bruffee quotes philosopher Karl Jaspers—"is bound up with communication."

Critique

Bruffee deserves praise for his promotion of a more student-active education through the method of collaborative learning-writing, and for his meditations on a theory of knowledge upon which to base that teaching method. I agree with the group-oriented nature of collaborative learning-writing, but there are problematic elements within the concept and its theoretical framework, particularly as they apply to the sciences and to the kinds of students who enroll in technical and professional communication courses. First, Bruffee's theory of knowing equates knowledge with "consensus," or in Rorty's terminology, with "socially justified belief," and thus collapses any difference between knowledge on the one hand and opinions and beliefs on the other. Why is this a problem? Because, as philosopher and historian of science Imre Lakatos argues, the differentiation between knowledge and belief has "vital social and political relevance." In *The Methodology of Scientific Research Programmes* Lakatos explains that philosophers have attempted to solve the problem of demarcation between knowledge and belief, between science and pseudo-science, by arguing that "a statement constitutes knowledge if sufficiently many people believe it sufficiently strongly"—in other words, if there is a "consensus" regarding the statement. But, Lakatos adds, history shows us that many people have been totally committed to absurd beliefs: "If the strength of beliefs were a hallmark of knowledge, we should have to rank some tales about demons, angels, devils, and of heaven and hell as knowledge" [8, p. 1]. People have been tortured and killed on the faulty basis of "socially justified beliefs"—in medieval Europe, in Nazi Germany, in apartheid-driven, white-enclaved South Africa. In itself, neither strength of commitment nor strength of numbers is a determinant of knowledge and/or truth. Lakatos continues: "a statement may be pseudo-scientific even if it is eminently 'plausible' and everybody believes in it, and it may be scientifically valuable even if it is unbelievable and nobody believes in it. A theory may even be of supreme scientific value even if no one understands it, let alone believes in it" [8, p. 1]. For Lakatos, then, knowledge is categorically different from belief and understanding; it is established or determined neither by a mental state (i.e., by believing) nor by discourse (i.e., agreement or "consensus" through discussion). Just which criteria do establish scientific knowledge in Lakatos' epistemology we shall turn to later; here our main point is that Bruffee's equation of "consensus" and "socially justified belief" with knowledge is, in the least, controversial and, in Lakatos' view, incompatible with a scientific attitude and the project of science.

Second, Bruffee accepts, falls into, a myth of revolution in his explanation of scientific change, progress, and shifts of knowledge bases within communities. He mystifies the patterned, accumulative process of change in the sciences through the use of terms such as "revolution" and "abnormal discourse." The metaphor of the scientific "revolution" equates change or progress with spectacular shifts of consciousness and knowledge, a notion in contradiction with the methods of scientific investigation and the accumulative process of scientific research programs. Lakatos critiques the notion of the "scientific revolution" as "irrational" in its connotation and unhistorical; nothing in the history of science, he contends, supports the paradigm of "revolutionary" change.

Similarly, Bruffee's concept of "abnormal discourse" is the metaphor of "revolution" in a different guise. Distinguishing between "normal" and "abnormal" discourse, Bruffee argues that the former maintains the established knowledge of a discourse community, while the latter, initially perceived by the community as "kooky" and "revolutionary," questions the beliefs of the community and thus paves the way for a reconstitution of knowledge [5, p. 648]. Both the metaphor of revolution and the notion of "abnormal discourse" imply that change is initiated by something aberrant, something outside of the usual discourse that characterizes the community. Through "revolution" and "abnormal discourse," Bruffee tries unsuccessfully to account for change in a model/system whose end-point is "consensus" or agreement. The concepts of "revolution" and "abnormal discourse" introduce disharmony into a model/system whose end-point is harmony, but as such they are inconsistent with that system and can only exist outside of it—that is, the one as unusual, spectacular change and the other as something "abnormal."

Third, Bruffee's theory of knowledge does not provide criteria for measuring the progress or degradation of knowledge and for evaluating rival claims to knowledge; within his theory, advancement is defined as shifts of consensus and movements from one discourse community to another. But without a standard of measuring the value of rival claims, the "knowledge" that one discourse community asserts is equally as good as the "knowledge" that the other asserts. Within Bruffee's model/system, justifying a belief involves nothing more than establishing "a certain kind of relationship among ourselves and among the things we say"; learning involves nothing more than leaving "the community that justifies certain beliefs" and joining "another community that justifies other beliefs" [9, p. 104]. Within such a model, "progress" and "advancement," terms which imply a scale of measurement, become vague "shifts" and "changes." For the sciences, Bruffee's model masks the fact of progress in areas such as the medical sciences, for example. We can meaningfully talk about "progress" or "advancement" in the treatment of certain diseases, rather than mere "shifts" or "changes" in methods of treatment. In regard to the sciences, then, Bruffee is locked into an epistemological position that our management of the world contradicts: for him, knowledge is merely interpretation, with one interpretation equal to another if it is consensual, socially justified.

Lakatos' "Methodology of Scientific Research Programmes"

Referring to Kuhn, Bruffee puts forward a discourse-bound criterion for determining the relative merit of competing scientific theories. He argues that scientists can reconcile conflicting, incompatible assumptions "only by debate," and in that debate, Bruffee explains, "scientists must rely on techniques that are neither 'straightforward, [nor] comfortable, [nor] part of the scientists' normal arsenal' " [9, p. 102].

But according to Lakatos, science does not operate in this way. Lakatos' "methodology of scientific research programmes" presents a non-discourse bound criterion for evaluating the relative merit of rival claims to knowledge: the criterion of "novel facts." In Lakatos' explanation of change in the sciences, progress takes place when a "scientific research programme," which consists of a body of related theories, meets three criteria. First, the theory "predicts some novel fact"; this constitutes "theoretical progress." Second, some of the "empirical content" of the theory is "corroborated by observation"; this constitutes "empirical progress." Third, "the hypothesis is consistent with the heuristic," or to say this in another way, the answer should be consistent with the question [10, p. 124]. Lakatos explains that a theory that does not meet these criteria represents in the first case "theoretical degeneration" and in the second, "empirical degeneration" [10, p. 124].

Thus, within this model of change in the sciences, progress is determined not by *consensus* but by the *ascendance* of progressive research programs, which predict more "novel facts" than their rival "degenerating" research programs. Lakatos prefers the word "progress" to "revolution"; alluding to Kuhn, he argues that "the rationale of scientific revolutions" lies in the advance of a research program (i.e., its prediction of new facts) and the degeneration of rival programs (i.e., their explanation of old facts but their inability to predict new ones). On this basis, scientists tend to join the progressive program and a consensus emerges. But the consensus itself does not establish the worth or the knowledge-content of one program in relation to another, and, Lakatos adds, it is not dishonest for scientists to stick with a degenerating program and try to turn it into a progressive one that will predict more new facts than its competitors [8, p. 6]. Taking aim at both Kuhn's and Popper's theories of scientific knowledge, Lakatos declares: "Kuhn is wrong in thinking that scientific revolutions are sudden, irrational changes in vision. The history of science refutes both Popper and Kuhn: on close inspection both Popperian crucial experiments [i.e., the idea that one key experiment can prove a theory] and Kuhnian revolutions turn out to be myths: what normally happens is that progressive research programmes replace degenerating ones" [8, p. 6]. Lakatos backs up this contention with lengthy discussions of the history of Copernicus's, Newton's, and Einstein's theories.

Lakatos' methodology of research programs puts forward a rational explanation of change in the sciences. It is "rational" in the sense that it offers criteria for measuring the strength of rival claims to knowledge and defines change or progress on the basis of those measurements. Lakatos' criteria are "objective" in that they refer to the natural world, the world of objects which also exist *outside* of discourse. Conversely, the concept of knowledge as "consensus" is "irrational" and debilitating because it offers neither criteria for measuring rival claims to knowledge nor a definition of change in the sciences that is not circumscribed by the subjective world of discourse, rhetoric, argumentation.

Lakatos' methodology incorporates the concept of consensus (that is, scientists tend to side with progressive research programs) without making consensus itself the factor which determines or establishes scientific advancement. Like Bruffee's theory of knowledge, Lakatos' methodology takes the double position that experience *in itself* cannot prove theoretical statements and that theoretical statements are by their nature fallible. Yet unlike Bruffee's theory, Lakatos' methodology, its combination of criteria, provides what philosopher Alex Callinicos calls a "highly sophisticated means of determining the relative merit of theories by virtue of their success in explaining the world" [10, p. 125].

Of course, scientists disseminate and promote their ideas through discourse, through publications and presentations, but, according to Lakatos, this is not the decisive element that establishes one idea or one aspect of a research program over another. That decisive element pertains to something that lies outside of discourse. Thus, within Lakatos' epistemology, there is a difference between objects and subjects, and between knowledge and belief; and science is grounded in this important difference.

To summarize this section, Bruffee's concept of collaborative learning-writing is based on a theory of knowing that equates knowledge with "consensus" and "socially justified belief"; it equates the humanities, which have discourse as their object of study, with the sciences, which have the natural world, the world which also exists outside of discourse as their object of study. Bruffee takes this epistemological position in order to set collaborative learning-writing on meaningful theoretical ground; as he says in *A Short Course*, if we conceive of science as "a process of gaining new information" and of knowledge as reflective of the natural world, then collaboration can only be "the blind leading the blind." In the first section of this chapter I have argued that teachers and theorists of technical and professional communication need a theory of knowing that refers to the world outside of discourse and that differentiates between knowledge and belief; in the second part I will argue that we can adopt such a theory and still conceive of collaborative learning-writing as a meaningful activity. To differentiate between knowledge and belief is not antithetical to a concept of collaborative learning-writing. At this point, we need to turn to the writings of the Bakhtin Circle, in particular to their differentiation between the natural and the human sciences, their distinction between knowledge and understanding, and their model of communication as collaboration.

THE BAKHTIN CIRCLE AND A THEORETICAL FOUNDATION FOR COLLABORATIVE LEARNING-WRITING

During the past decade, the writings of the Bakhtin Circle have had an expanding influence on linguistics, speech communication, and literary studies in the United States; these writings also constitute a potentially rich field of investigation for theorists of composition and technical and professional communication.[2] In the second section of this chapter we will discuss those aspects of the Circle's writings that are germane to a theoretical foundation of a concept of collaborative learning-writing. Those aspects are: 1) the demarcation between the natural and human sciences and the concomitant differentiation between knowledge and understanding, 2) a theory of the socially determined nature of language, signs, and communication, 3) a model of communication as collaboration, and 4) a philosophy whose central concept is "dialogism," the interaction between self and "other" through which identity and meaning are created. Bakhtin's ideas strengthen the weaknesses within Bruffee's theoretical foundation for collaborative learning-writing and provide additional support for those already strong aspects of Bruffee's concept.

The Differentiation of Knowledge and Understanding

The differentiation of knowledge and understanding in Bakhtin's epistemology derives from his differentiation of the natural and the human sciences; this latter division itself depends upon divisions between "subjects" and "objects," between "monologue" and "dialogue," and between "primary datum" and "ideology." We will need more than a few paragraphs to explain these terms and concepts because Bakhtin gives a special signification to them and because important points that we will make about revising Bruffee's concept of collaborative learning-writing will depend on Bakhtin's differentiation of knowledge and understanding.

Bakhtin argues that the "natural sciences" and the "human sciences" have different foci of study: the former study *objects*; the latter, *subjects* or subjects' *discourse* of one kind or another. The human sciences comprise those disciplines which have "texts," written or oral, as their data, their "primary given" [14, p. 105, 107]. By "texts" Bakhtin means thoughts, speech, and actions, all of

---

[2] The term "Bakhtin Circle" refers to a group of Soviet intellectuals, artists, and scientists who were influenced by and collaborated with literary theorist, linguist, and philosopher, M. M. Bakhtin. This chapter cites works of three members of the group: Bakhtin, linguist and semiotician [11–13], V. N. Voloshinov, and literary theorist P. N. Medvedev. The authorship of certain texts of the Bakhtin Circle remains controversial: I cite two of these texts: *Marxism and the Philosophy of Language* (1929), published under the name of V. N. Voloshinov, and *The Formal Method of Literary Scholarship* (1928), published under the name of P. N. Medvedev—a slash between names, as in "Voloshinov/ Bakhtin," indicates the controversial, collaborative nature of the authorship of these texts.

which must be "read" or "interpreted"; texts can be understood only in the "dialogic context" of their time, that is, as "rejoinder(s)," as "semantic position(s), or as "system(s) of motives." In short, texts are always responses to something. The human sciences do not concern themselves with Man as object,[3] but solely with Man as a participant within or as a maker of texts [14, p. 105, 107]. One major difference between the human and natural sciences, then, is that the former study "subjects," "texts," or discourses, while the latter study "objects," "voiceless thing(s)."

In Bakhtin's division, the natural and human sciences have different ends as well as different foci: the end of the former is *knowledge,* whereas the end of the latter is *understanding.* Because natural sciences study objects that "communicate nothing of themselves," these sciences produce a knowledge that has nothing to do with "discourses or signs." Conversely, the human sciences study all the various problems of "establishing, transmitting, and interpreting the discourse of others" [11, p. 15]. In sum, the natural sciences seek to "know" objects, whereas the human sciences seek to "understand" signs, texts, discourses.

Bakhtin elaborates on the difference between "knowledge" and "understanding" through the distinction between two more terms, the "monologic" and the "dialogic." In the natural sciences, Bakhtin explains, "the intellect contemplates a *thing* and expounds upon it"; this is "monologic" in the sense that there is *one* subject in opposition to a mute object. Conversely, in the human sciences, an intellect contemplates another intellect; this is "dialogic" in the sense that there are *two* subjects in exchange [12, p. 161]. Or to say this in another way, the human sciences contemplate society (a multitude of subjects who constantly enter into discourse with one another), while the natural sciences study "voiceless things" that exist outside of society [12, p. 161]. For Bakhtin, nature also exists outside of society, and knowledge pertains to this (outside) "monologic" world of "physical and chemical bodies." Conversely, human "texts," "the products of ideological creation," can develop only within and for society; the "dialogic" enterprise of the humanities is to understand these products [13, p. 17].[4]

For Bakhtin, understanding is active, not passive: this is perhaps the most important point to grasp about the concept, and the difference between Bakhtin's use of this word and our own. For most of us think of understanding as still, passive, meditative, rather than something active, struggling, something carrying within itself a notion of conflict, resolution, and renewed conflict

---

[3] Bakhtin writes that it is possible to study man as an object, but that when man is studied "outside of a text and independent of it, the science is no longer one of the human sciences . . ." [13, p. 105].

[4] Here we should note that, for Bakhtin, "ideology" is not, as it is for us, a political buzz-word; it means "idea-system," and pertains to the "concrete exchange of signs in society and in history" [4, p. 429]. Human speech, actions, and aesthetic creations such as literature, art, and music are ideological in the sense that they embody the patternings unique to their respective historical, sociological milieux.

depending upon the depth of the understanding. V. N. Voloshinov, one of Bakhtin's collaborators, writes that understanding orients the listener/reader in relation to a particular "utterance" (or discourse) and places it in its proper "context." Understanding constitutes "the germ of a response." That is, understanding "lay(s) down a set of . . . answering words" to match the words of the utterance; it "strives to match the speaker's word with a *counter word*" [3, p. 102]. In this sense, the listener/reader becomes the speaker/writer [2, p. 68]. Thus, understanding is "dialogic" in that it involves at least two intellects (subjects) in conjunction, in collaboration.

Understanding is "dialogic," both an individual and a social act. Voloshinov writes that within each utterance a "dialectical synthesis" occurs between the individual and society, between the "inner" and the "outer." He explains, brilliantly: "In each speech act, subjective experience perishes in the objective fact of the enunciated word-utterance, and the enunciated word is subjectified in the act of responsive understanding in order to generate, sooner or later, a counter statement" [3, p. 41]. Voloshinov calls the words of each utterance "little arena(s) for the clash and criss-crossing of the differently oriented social accent" [3, p. 41]. He connects the individual and his or her utterance with society today (which, as we shall see, also bears the traces of yesterday's society): "A word in the mouth of a particular person is a product of the living interaction of social forces" [3, p. 41].

Inherent in this concept of understanding as an individual yet social act is a theory of language, sign, and communication as socially determined phenomena. It is to a more detailed explanation of this theory that we now turn.

The Uniqueness of the Utterance and the Social Nature of Language and Sign

Interest in collaborative learning-writing has fueled the debate between two groups of composition and writing theorists: those like Peter Elbow who emphasize the individuality of the speaker/writer, and those like Kenneth Bruffee who emphasize the social shaping of the speaker/writer. Voloshinov/Bakhtin's theory of language and sign can shed some light on this debate.

In *Marxism and the Philosophy of Language* (1929) Voloshinov summarizes two trends of thought on language which he terms "abstract objectivism" and "individualistic subjectivism." The former trend focuses on "the system of language" (on *langue*) and dismisses the utterance, or the speech act, because it views the utterance as individual and therefore not sufficiently interesting in terms of the overall, abstract language system. The latter trend focuses on the utterance (on *parole*) and explains language in terms of the "individual psychic life of the speaker" [3, p. 82]. Voloshinov considers neither of these trends or orientations correct; he synthesizes positive aspects of them into a third orientation that focuses on the utterance yet does not view it as wholly determined by "the physic life of the speaker."

Voloshinov considers the individual utterance as unique, in one sense, yet thoroughly social in another. First, he defines language as utterance; he conceives of language as *"a continuous generative process implemented in the social-verbal interaction of speakers"* [3, p. 98]. Language changes constantly; therefore, there is always something new, and old, about it. Each utterance, each exchange between speaker and listener, between writer and reader, is unique, different from any exchange before or after; by this Voloshinov does not mean the speaker's or listener's words are uniquely his or hers alone, but that their interaction, collaboration, at this particular moment, in this particular context, in this particular utterance differs from preceding and succeeding utterances. Secondly, Voloshinov considers the laws of language and the *"structure of the utterance"* as *"purely sociological"* [3, p. 98]. He views the utterance as an interplay between "context" and "content." Bakhtin writes that the context, "like the sculptor's chisel, hews out the rough outlines" of the utterance; the context concentrates and focuses the utterance's "internal impulse"; its content [4, p. 357-358].

So, to return to the question we posed indirectly at the opening of this section, what is unique about the utterance is the conjunction of a particular writer, the intended reader(s), and their linguistic engagement shaped by the particular context. This conjunction will never be the same again, because it will be enriched by other discourse and altered by a changing context. Each utterance is unique in its particular, temporal conjunction, yet, at the same time, each utterance is thoroughly, completely, socially determined, socially patterned. In Voloshinov/Bakhtin's theory of language, the individual's "inner world" has a "stabilized *social audience* that comprises the environment in which reasons, motives, values, and so on are fashioned" [3, p. 86]. It is to this inner, stabilized social audience that Voloshinov refers when he states: *"The immediate social situation and broader social milieu wholly determine— and determine from within—the structure of an utterance"* [3, p. 86]. Individual expression is always a socialized expression: "The stylistic shaping of an utterance is shaping of a social kind, and the very verbal stream of utterances, which is what the reality of language actually amounts to, is a social stream. Each drop of that stream and the entire dynamics of its generation is social" [3, pp. 93-94]. Bakhtin/Voloshinov's theory of language is at its base a theory of signs: "Every sign . . . is a construct between socially organized persons in the process of their interaction. Therefore, *the forms of sign are conditioned above all by the social organization of the participants involved and also by the immediate conditions of their interaction*" [3, p. 21].

We can perceive in the Bakhtin Circle's theory of sign and language the elements of a conception or model of communication as collaboration. Each utterance is unique, yet because the utterance is socially determined, speakers and writers constantly encounter the words of "others," whose "voices" can be heard, faintly or forcefully, in everything that a person says or writes. Behind the burst of interest in collaborative learning-writing lies an increasing sensitivity to the utterance and to the complex collection of social forces that bear upon it.

Communication As Collaboration

We are all familiar with the standard diagram of the process of communication. This model comprises a "sender" who "encodes" and sends a "message" that travels through a "channel/medium" and a "receiver" who "decodes" the message and transmits "feedback" to the sender [15, pp. 16-18]. There is very little that is collaborative about such a model of communication. In opposition to this model, the Bakhtin Circle represents communication as a process in which meaning is constructed in the interaction between speech-act participants. Explaining the differences between these two models of communication, Medvedev/Bakhtin emphasize three points. First, the relationship between speaker/writer and listener/reader changes constantly in the process of the communication. Second, there is "no ready-made" message, but rather, one which is itself "generated" in the interaction between speaker/writer and listener/reader. And third, this message is not "transmitted," but "is constructed between them as an "ideological bridge" built "in the process of their interaction" [16, p. 152].

Communication is thus an inherently collaborative process which involves the building of "ideological bridge[s]" between speaker/writer and listener/reader. Using a related metaphor, Voloshinov compares "meaning"—that which is generated in the "interaction" between speaker/writer and listener/reader—to an "electric spark that occurs only when two different terminals are hooked together" [3, pp. 102-103]. These two metaphors—the process of bridge building, the process of the spark generated by linked electrical terminals—depict a process in which speech-act participants are mutually active, mutually contributors to and makers of the meaning of the communication. In the context of the Bakhtin/Medvedev metaphor, the word "ideological" denotes not something chiefly political, but rather the complex spectrum of ideas, from fragments to systems, which characterize a particular society [17]. In Bakhtin/Voloshinov's view, ideological patternings take place both within and outside of the individual; the individual psyche is "thoroughly ideological," thoroughly "sociological" [18, p. 31]. But a certain receptiveness, a pre-patterning, must exist before bridge-building between speech-act participants can take place. Present attitudes, beliefs, past behavior—all of which are both individual and social—influence a communication. As Voloshinov explains, there is no absolute neutrality in the realm of discourse: "we never say or hear *words*, we say and hear what is true or false, good or bad, important or unimportant, pleasant or unpleasant, and so on. *Words are always filled with content and meaning drawn from behavior or ideology* ... and we can respond only to words that engage us behaviorally or ideologically" [3, p. 70]. Furthermore, because past utterances do not "die," but reflect, even shape, present utterances, speaker/writer and listener/reader do not face each other like "sovereign egos" across an "uncluttered space," but join or collaborate with other interlocuters: "A single voice can make itself heard only by blending into the complex choir or other voices already in place" [4, p. XX; 22, p. x]. The human space within and about us is steeped in past and present discourse, and we try to communicate, collaboratively with others, in that "cluttered" space.

Bakhtin extends the dialogic principle, the essence of his model of collaborative communication, into a worldview that he terms "philosophical anthropology" [1, p. 146]. This philosophy contends that "others"—both objects (like the sun and the moon) and subjects (like friends and colleagues)—contribute to the construction of our identity and the making of our meaning of the world: ". . . we appraise ourselves from the point of view of others, we attempt to understand the transgredient moments of our very consciousness and to take them into account through the other . . . in a word, constantly and intensely, we oversee and apprehend the reflections of our life in the plane of consciousness of other men" [19, p. 94]. For Bakhtin, the model of communication and the philosophy become the same: "Life by its very nature is dialogic. . . . To live means to participate in dialogue: to ask questions, to heed, to respond, to agree and so forth" [20, p. 293]. In sum, for Bakhtin "*To be* means to *communicate*" [20, p. 287]. Life, as well as communication, is collaboration—a philosophical view that recalls the lines of Robert Frost which Kenneth Bruffee is fond of quoting: "Men work together, . . . / Whether they work together or apart" [6, p. 334].

Conflict Within Collaboration

Michael Holquist, one of Bakhtin's English-language translators and his biographer, neatly sums up the *a priori* of Bakhtin's philosophy as "an almost Manichean sense of opposition and struggle at the heart of existence." Holquist writes that for Bakhtin this struggle is present in the individual consciousness, in the utterance, in culture and society, in nature [4, p. xviii].

Holquist's interpretation seems to me particularly germane to a differentiation between Bruffee's and Bakhtin's concept of communication. Many similarities exist between the two, but they are different in at least one important respect: the terms "consensus" and "dialogism" suggest essentially different processes of interaction between speaker/writer and listener/reader. Words such as "dialogue," "dialogism," and "dialectic" denote an interaction where two opposing forces meet, where a counter-response matches a response. A sense of difference, of otherness within the interaction, is implied in these words. Conversely, Bruffee's concept of collaboration de-emphasizes the idea of struggle or conflict; in fact, in the "Introduction" to *A Short Course* Bruffee recommends collaborative learning-writing as an antidote to an overly competitive, hyperactive concept of conflict and struggle which views "others" as the enemy, the adversary, the ones to be overwhelmed by argument [6, pp. 2-3]. For Bruffee, collaboration and communication are a form of social bonding; Bruffee says that we write in order "to maintain our membership in communities . . ., to make ourselves acceptable to communities we are not yet members of, to help other people join communities that we are already members of, or to define communities that we believe we and our readers are already members of . . ." [6, p. 3]. Collaborative learning-writing works toward "consensus," the "join[ing] of assenting communities of knowledgeable peers"; but with this emphasis on assent and agreement, the theory of knowledge underlying collaborative learning can accommodate difference only by introducing an additional, extraneous concept, that of "abnormal discourse," which as it turns out in Bruffee's theory, plays a key if not *the* key role in the genesis of shifts of knowledge in discourse communities. Bruffee's theory of knowledge as consensus privileges agreement, but according to his model, shifts of knowledge in discourse communities seem to depend, ironically, on "abnormal discourse" or disagreement. We can easily imagine situations—in fact, we are constantly involved in them—in which the difference and disagreement of others, of our collaborators, pose questions demanding counter-responses, a carrying of the discourse to a deeper level of understanding.

For Bakhtin, the end of communication is understanding, not consensus, and in that act of understanding "a struggle occurs that results in mutual change and enrichment" [21, p. 142]. In Bakhtin's model of communication, the utterance comprises response and counter-response(s); communication requires ideological bridge-building, yet that bridge-building, a metaphorical spanning of two different points, involves difference, otherness, struggle. Bruffee's concept of consensus, a collapsing of difference, seems to end in a communal stasis that must wait to be jarred out of its slumber by "abnormal" discourse. But what is the source of this abnormality? Does it exist latently within the community, or does it creep in from outside?

Bakhtin conceives of communication as utterances interlinked without end: "The word wants to be heard, understood, responded to, and again, to respond to the response, and so forth *ad infinitum* . . ." [14, p. 127]. Rather than "consensus" or social bonding, the goal of communication in Bakhtin's model is ever greater "*depth* of understanding" [14, p. 127]. Perhaps there is a suggestion in this notion of "depth" that understanding for Bakhtin moves toward something that we might call not knowledge, but truth, an ever more profound creating of identities and meaning through our collaboration with each other. In Bakhtin's view, this involves infinite struggle.

CONCLUSION

The Bakhtin Circle's writings constitute a rich source of interrelated, complex ideas about the collaborative nature of human communication. In general, their writings provide support for Bruffee's concept of collaborative learning-writing.[5] In particular, they amplify key ideas and shore up weaknesses in the theory of knowledge on which Bruffee's concept is built. First, because Bakhtin establishes a demarcation between human and natural sciences, and between understanding (which pertains to "texts") and knowledge (which pertains to the natural world that exists outside of texts), Bakhtin maintains a division between objectivity (having to do with "objects") and subjectivity (having to do with "subjects"), and between the "ideological" world of texts and the non-ideological, natural world which texts reflect and refract but do not bind. Through the natural sciences, we know the natural world; through the human sciences, which are based on sign, texts, discourse, we learn to understand ourselves in the human, social world. Thus, the goal of collaborative learning-writing in the composition and technical and professional communication classroom is not, as Bruffee contends, knowledge, but understanding, as that understanding applies to the human "text" in the broadest, richest sense of the word. Second, the Bakhtin Circle views communication as a collaborative process in which speech-act participants are equally speakers and listeners who engage in ideological bridge-building, a making of identity and meaning through and within the speech act. But this collaboration also involves difference and struggle, which can lead to a more profound cooperation, understanding. Third, for the Bakhtin Circle sign and language are socially-determined phenomena; individual speech-act participants make meaning in each, unique utterance, but the meaning that they make is socially determined, shaped by utterances that have preceded it and those that will follow it. Fourth, Bakhtin's model of communication as dialogic exchange extends to a "philosophical anthropology" in which human, social life is defined as utterance, as dialogue, as communication.

Finally, this chapter has recommended three revisions and/or clarifications of Bruffee's concept of collaborative learning-writing and the theory underlying it. First, the collaboration should be based on a theory of knowledge that maintains a division between the sciences and the humanities, and knowledge and beliefs. Second, collaborative learning-writing has as its end, or goal, understanding, not establishing or generating knowledge. Among other things, this second revision/ clarification will render collaborative learning-writing more acceptable to our scientific, technological-minded students, who do distinguish between knowing and believing, and between data pertaining to the natural world and the rhetorical world of discourse. Third, collaboration is not a monologic movement toward a sameness or a "consensus," but a dialogic movement in which difference and struggle are inherently part of the bridge-building of understanding.

---

[5] In the Introduction to *The Dialogic Imagination,* Michael Holquist points out that Bakhtin's thought parallels Vygotsky's, a sociolinguist who has influenced Bruffee's thinking on communication [4, p. XX].

REFERENCES

1. M. M. Bakhtin, cited in T. Todorov, *Mikhail Bakhtin: The Dialogical Principle*, Wlad Godzich (trans.), University of Minnesota Press, Minneapolis, 1984.
2. _____, The Problem of Speech Genres, *Speech Genres and Other Late Essays*, V. W. McGee (trans.), C. Emerson and M. Holquist (eds.), University of Texas Press, Austin, pp. 60–102, 1986.
3. V. N. Volos(h)inov, *Marxism and the Philosophy of Language*, L. M. Matejka and I. R. Titunik (trans.), Harvard University Press, Cambridge, Massachusetts, 1986.
4. M. M. Bakhtin, *The Dialogic Imagination*, C. Emerson and M. Holquist (trans.), M. Holquist (ed.), University of Texas Press, Austin, Texas, 1981.
5. K. A. Bruffee, Collaborative Learning and the 'Conversation of Mankind,' *College English, 46*:7, pp. 635–652, 1984.
6. _____, *A Short Course in Writing: Practical Rhetoric for Teaching Composition through Collaborative Learning*, 3rd Edition, Little, Brown and Company, Boston, Massachusetts, 1985.
7. _____, The Structure of Knowledge and the Future of Liberal Education, *Liberal Education, 67*:1, pp. 177–186, 1981.
8. I. Lakatos, *The Methodology of Scientific Research Programmes*, Vol. 1, J. Worrall and G. Currie (eds.), Cambridge University Press, Cambridge, England, pp. 1–7, 1978.
9. K. A. Bruffee, Liberal Education and the Social Justification of Belief, *Liberal Education, 68*:2, pp. 95–114, 1982.
10. A. Callinicos, *Marxism and Philosophy*, Oxford University Press, Oxford, 1985.
11. M. M. Bakhtin, Discourse in the Novel, cited in T. Todorov, *Mikhail Bakhtin: The Dialogical Principle*, Wlad Godzich (trans.), University of Minnesota Press, Minneapolis, 1984.
12. _____, Toward a Methodology for the Human Sciences, *Speech Genres and Other Late Essays*, V. W. McGee (trans.), C. Emerson and M. Holquist (eds.), University of Texas Press, Austin, pp. 159–172, 1986.
13. _____, The Problem of the Text in Linguistics, Philology, and the Other Human Sciences: An Essay on Philosophical Analysis, cited in T. Todorov, *Mikhail Bakhtin: The Dialogical Principle*, Wlad Godzich (trans.), University of Minnesota Press, Minneapolis, 1984.
14. _____, The Problem of the Text in Linguistics, Philology, and the Human Sciences: An Experiment in Philosophical Analysis, *Speech Genres and Other Late Essays*, V. W. McGee (trans.), C. Emerson and M. Holquist (eds.), University of Texas Press, Austin, pp. 103–132. (This is a different translation of the essay cited in 13.)
15. H. A. Murphy and H. W. Hildebrandt, *Effective Business Communications*, 4th Edition, McGraw-Hill Book Company, New York, 1984.
16. M. M. Bakhtin and P. N. Medvedev, *The Formal Method in Literary Scholarship: A Critical Introduction to Sociological Poetics*, A. J. Wehrle (trans.), Harvard University Press, Cambridge, Massachusetts, 1985.

17. G. Therborn, *The Ideology of Power and the Power of Ideology*, NLB, London, 1980.
18. V. N. Volos(h)inov, *Freudianism: A Marxist Critique*, I. R. Titunik (trans.), Academic Press, New York, 1976, cited in T. Todorov, *Mikhail Bakhtin: The Dialogical Principle*, Wlad Godzich (trans.), University of Minnesota Press, Minneapolis, 1984.
19. M. M. Bakhtin, "Author and Character in Aesthetic Activity," cited in T. Todorov, *Mikhail Bakhtin: The Dialogical Principle*, Wlad Godzich (trans.), University of Minnesota Press, Minneapolis, 1984.
20. M. M. Bakhtin, *Problems of Dostoevsky's Poetics*, C. Emerson (ed. and trans.), University of Minnesota Press, Minneapolis, 1984.
21. M. M. Bakhtin, From Notes Made in 1970–71, *Speech Genres and Other Late Essays*, V. W. McGee (trans.), C. Emerson and M. Holquist (eds.), University of Texas Press, Austin, pp. 132–159, 1986.
22. Introduction to T. Todorov, *Mikhail Bakhtin: The Dialogical Principle*, Wlad Godzich (trans.), University of Minnesota Press, Minneapolis, 1984.

CHAPTER 3

The Construction of Multi-Authored Texts in One Laboratory Setting

JAMES R. WEBER

The process approach is a particularly suitable way of examining group writing because it encourages observation of group interactions as well as individual writer's actions that directly involve invention and revision. However, in the most-used model of the writing process, the Flower and Hayes model, the writing environment is a nebulous influence on the individual writer's experience [1]. This vagueness is not surprising because the model is primarily intended to show the individual's sequence of cognitive acts (planning, translating, and reviewing) in processing information and finding suitable presentational forms. Feeding this cognitive process are the individual writer's memories, scripts (internalized frameworks of professional practice), concepts, and current goals, and, more vaguely, the "task environment" that offers the individual writer constraints and mandates.

However, accounting for the task environment is at the heart of developing a theory of the collaborative writing process because an individual author's environment is defined by a past or present relationship to other authors. The social constructionist approach to collaborative writing has addressed the formative qualities of the task environment. Those scholars using a social constructionist approach regard knowledge as socially formed; thus, research into the construction of scientific texts has seen them as reflections of the professional communities in which they are produced [2-5]. Through text-based interviews, research in professional writing has tracked the social bases of writers' decisions about wording, format, tone, and other content to writers' motivations to confirm their memberships in particular social communities [6-9]. These researchers hold that texts are social "constructions" rather than organic outgrowths of cognition.

The theoretical approach to multi-authored writing, however, must not be hung up on either its social or cognitive aspects. To overemphasize the significance of either approach is to ignore the significance of the other. Any serious theory of collaborative practice must be integrative, taking into account the writing process of the individual and the group. As a contribution to such a theory, this study addresses a preliminary question: What are the elements in the writing environment that most strongly influence the individuals in the co-authoring process?

In addressing this question, I have assumed that professionals form frame-works of beliefs and information that allow them to apply and develop their trade knowledge as they practice. Thus, I have sought areas of structured communal experience from which authors establish and then maintain a common working framework, one which would incorporate knowledge of the organizational structure, the dynamics of interactions with other authors, and their personal writing process.

As a technical editor in a large research laboratory (Battelle Northwest), I have both worked with and observed scientists as they coauthor works with different rhetorical properties—varying purposes, audiences, and content. There appear to be distinctive elements in this task environment that recur frequently as authors talk about planning, writing, and reviewing their work. To locate these elements and name them, I interviewed six authors in a single department over a period of six months, while also following the process of their work from planning through publication. Wanting to know both the individual and the collective experience, I gathered information about features of the writing environment that are shared.

From a cognitive standpoint, it was soon evident that these authors share, probably unknown to each other, a vision of the ideal process of coauthoring (see [10] for a similar observation). They also share the uncertainties that writers, in their roles as decision-makers, may experience during the process of writing: uncertainties about the choice and ordering of information, about appropriateness for the intended audience, perhaps even about the identity and interests of their audiences. Besides the uncertainties that individual writers experience, these writers also had potential uncertainties associated with co-operating with other authors: the possibilities of divergent intentions and multiple interpretations. As studies of business communication have also shown, these coauthors sought consensus on matters relating not only to the substance of the documents produced but also on the methods of producing them [11].

Moreover, their approaches to coauthoring involve a knowledge of their organizational environment. Scientists who were new to the organization had to learn the organizational lore, such as requirements of the document-cycling process, even though they may have written reports and collaborated with other scientists before they came to Battelle.

From interviews and observations, I found three overlapping areas of knowledge from which frameworks could be refined: the protocols of document leadership and authorship; the determination of audience, goals, and content for coauthored documents; and the resolution or avoidance of disagreement among coauthors. These three areas of knowledge cut across the cognitive and social constructions of texts; individual authors working through their uncertainties and groups working toward consensus are processes that are profoundly intertwined. In addition, the organizational environment extends the coauthoring process beyond individual experiences of the coauthors; texts are constructed from individual and group writing processes, which are supported and mediated by the organizational review. Consensus in this environment is perhaps best thought of as "constructed" rather than "reached."

LABORATORY WRITING ENVIRONMENT

The laboratory department, which contains the authors studied, is a highly structured organization, with a well established document-cycling process, influential and actively involved sponsors of the work, and authors who are often physically separated from one another in various buildings or even in differing parts of the country.

The department is divided into three major sections encompassing ten technical groups. Overall, there are 151 persons working in the department, including 47 support staff (secretaries, word processors, clerks, and editors) and 104 technical staff including staff scientists and managers, who are drawn exclusively from staff scientists. Managers consist of a general department manager, two deputy managers who interact directly with the governmental and private clients, and managers over each of the department's sections. In addition, members of the scientific staff may be appointed ad hoc project managers for the duration of projects. The scientific staff who write reports and documents for sponsors report directly to their project managers.

The physical disposition of the department—the buildings in which they work and the layout of those buildings—forms much of the flow of daily communication among the scientific and management staff, who do most of the writing in the department. The department is scattered among a half-dozen buildings spread several miles apart from one another, with the various buildings housing similar technical groups or department sections. Most of the department is housed in a single building that contains nearly all the support staff; however, one section has offices in a building with no laboratory space, several miles distant from the main building; other sections work in laboratories closer to the main building but still require staff to travel by car, bicycle, or foot between buildings and often to spend considerable time on the telephone if they are working on projects with staff from other buildings. Indeed, because of occasional section reorganizations, some staff may become physically separated from their sections when most of their section is moved to another location.

Most of the department's clientele are government agencies, with one federal government agency providing approximately 60 percent of the work and a second providing 20 percent. Private contractors and state agencies provide the rest. Scientific staff and managers are responsible for generating proposals for work, answering government requests for proposals, and approaching private contractors. Thus, in addition to performing regular laboratory work, staff scientists are encouraged to seek new sources of funding.

Writing projects are related almost exclusively to sources of funding. The main product of the department is information rather than any physical object. Thus, written reports represent the outcome of work and satisfy the terms of the contracts engaged. Written products also generate the contracts and are necessary to clarify misunderstandings, resolve difficulties, and track the progress of work. Written work produced by the department can be distinguished into work meant only for readers internal to the laboratory and that meant for external readership, such as sponsors or the scientific community at large through conference presentations, articles, and books.

Because so much depends upon the documentation of work, scientific staff devote most of their writing efforts to producing sponsor reports. More than the other written products of the department, sponsor reports are heavily reviewed internally and externally. Sponsors, both government and private, may review multiple drafts of each report, suggesting changes in the text. Formal reports produced in the department go through typical revision cycles (see Figure 1). Anyone in the cycle (sponsor, manager, author, editor) has multiple opportunities to review the draft. Although more formal, this cycling is similar to that described by Paradis, Dobrin, and Miller at Exxon ITD [12]. Some of these cycles, such as the sponsor-comment cycles, are required by sponsors; others, such as the management- and peer-review cycles, are required by the internal clearance process at the laboratory. Indeed, each document intended for external distribution must be cleared through internal channels to prevent the dispersal of proprietary or classified information. Moreover, because the laboratory performs most of its work for the federal government, documents involving government projects must be cleared not only internally by the laboratory but also by federal government clearance. Although there are few changes in the text of documents by the external clearance authorities once a draft leaves the department, most authors speak of "clearance" as including internal department approvals (peer reviewing and management approvals) as

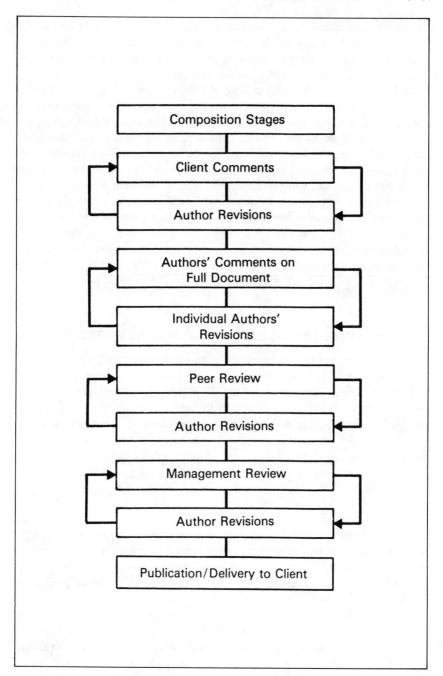

Figure 1. Document Cycling in One Battelle Laboratory Department

well as external clearance. Participants in this study all held a strong belief in the necessity and efficacy of the cycling process. Management signatures on the clearance form indicate mutuality—that one's own group of authors is not alone in standing behind the information in the document.

At least three aspects of this writing environment regularly influence authors' group writing processes. First, the separation of authors may compound the effects of multiple minds in authorship; thus, uncertainties may arise, particularly in large projects, about the aims or coherence of the eventual product. Where spontaneous consensus is not realistic, this potential uncertainty raises the need for a single guiding view of the writing process—e.g., a document leader—to ensure completion and coherence. Second, writers' communal understanding of the goals and content of the document can be influenced by a host of factors in the writing environment, foremost among them being sponsors' and managers' expectations. Third, authors need methods for resolving disagreements and inconsistencies among the contributors, a capacity undertaken both formally (through document cycling) and informally (in the halls, on the telephone, and via electronic mail). The following sections discuss these overlapping areas.

PROTOCOLS OF LEADERSHIP AND AUTHORSHIP

Group writing in this laboratory environment is rarely "collaborative" in the sense that Wiener [13] or Allen et al. [14] have used it: that is, individual contributors sharing power in making decisions that can be accepted by the group as a whole, "a situation in which decisions are made by consensus" [14, pp. 72, 73]. Such a situation is hardly possible when work must be reviewed by several levels of responsible persons, both contributors to the project and reviewers of it. Although writers may strive for such collaborative working arrangements, the dispersed working environment and organizational constraints lead them to settle for something other than decision-making shared among the authors. In fact, decisions about the content and goals of written work can be revised by reviewers, managers, or sponsors. By this definition of collaboration, then, those involved in the pathways of review and approval can be said to collaborate in the production of a document. Thus, authorship can become extended beyond the core of writers who compiled the drafts. Of course, this approach does not imply strict consensus. Dissenting opinions are possible, as we shall see below, and compromise may be needed.

Writers in this environment may work alone and then present their work to a lead author, project manager, or designated coordinator for incorporation into large documents. Like the department itself, which is spread among several buildings, the documents produced often involve authors spread widely apart, some in different corners of the nation. In this environment, sponsors and project managers may have major roles in controlling not only what the document says but how it is produced. Cooperation that allows a degree of collaboration on the content appears to be more possible, however, on smaller documents with few authors and on projects generated from the authors themselves rather than received from outside sponsors. Differences also exist in the working relationships within sections of the department, with some sections allowing more leeway in sharing partially completed drafts with other authors and other sections insisting on more completed drafts.

The participants in the study have often chosen to turn the interviews to the topic of leadership in group writing situations. Their concern with leadership reflects their need for reducing their uncertainties: for clarity about the project's purposes and about the scope and adequacy of their own contributions. In a tightly structured organization, in which many reviewers may change even a single-authored document, multi-authored works may raise issues of authority and responsibility. From the interviews, two types of document leaders emerge: lead authors and document coordinators.

A lead author was described as someone who has the authority to make changes in other authors' contributions and also has the authority to meet contractual deadlines, negotiate directly with the sponsor of the project, and either serve as project manager or represent the other authors to the project manager. The qualities of a lead, or first, author begin to sound somewhat idealized in some interviews, reflecting the importance that writers in this environment give to strong leadership in multi-authored works: "the person who does the most work, who generates the paper and puts it into words is first author;" "somebody who is in charge of the quality of a report, and that's the first author;" "responsible for every aspect of it;" "responsible for the overall scientific quality of the work and of the writing." The primary distinguishing trait of strong lead authors appears to be that they provide the other authors with a clear vision of acceptable content in the document through clarity about the purpose of the work and through a continuing attention to each contributor's progress.

The document coordinator oversees the work of other authors and collects their contributions. Coordinators rarely rewrite or attempt to fill in the gaps between texts from various authors. One writer explained that "in some cases, the project manager will not assign a lead author, but have someone who is supposed to 'shepherd it through.' In that case, they just look at sections. So, it's not until you have the final product that someone reads it from cover to cover." Often, coordinators are appointed by managers to track a work through the writing and production process, as other authors write the text. They may also be expected to provide guidance for authors by referring to the general goals of the work. Coordinators may also be listed as authors. However, the coordinator may also be the project manager, who has decided to intervene little in the writing, choosing instead to concentrate on handling the funding and running interference with the sponsor.

When problems arise in this environment, it may be because there is inadequate communication among coauthors. A possible weakness of coordinated documents is that coauthors may lack an involved, active coordinator to function as a clearinghouse of information. When authors are vague about the goals of the document or the nature of the content they are expected to contribute, they look to lead authors or coordinators to reduce that uncertainty.

Document leaders are usually appointed by project managers. Lead authors are often also the project managers if they are the principal investigators. As such, they also are responsible for negotiating with sponsors on deadlines and the suitability of content. Indeed, depending on management mandates, one scientist noted, he was either a lead author or a coordinator. On one project, he said:

> I did the conclusions, the summary, and the introduction. I filled in those spots that are general. Plus I also did several chapters myself. I was in charge of making sure that the budget was coming along okay. That we weren't overspending here or there. I called meetings to make sure we were on track and stuff. If we had to change schedule a little bit, I did that. I interfaced with the sponsor, who was going to receive this and who we were really writing it for.

On another project, as coordinator this time, he took finished copy from two authors and performed as a surrogate project manager:

> I take what they get, and I go to the editor with it, and to graphics and to word processing. I try to keep things on schedule. I remind the authors when things are due, and I talk to them about things, but I don't really set down and rewrite anything. . . . My manager said, these guys are to be coordinated. But it wasn't clear and I didn't interpret it to mean that I could rewrite what they wrote. I can make comments clearly, I can go back and say, I thought we agreed we were going to do it this way. The difference is that I didn't take their input and type it into the draft. Whereas, when I'm the lead author, I take the input and retype it into a final document, so I can take the liberty to make changes.

One author noted that lead authorship "wouldn't work for a large document because it's so time-consuming." Rethinking each chapter in a large project could be prohibitively expensive, perhaps duplicating the work of peer reviewers and editors, who will have a role in the final product in any case. Another reason is that some of the department's projects require authors with widely different kinds of expertise. "She does oranges, he does apples," one study participant noted about the strengths of each contributor, concluding that a good coordinator would recognize strengths and leave the writing tasks to the experts in each field.

The desire for a strong lead author, then, might be seen as the drive for an overarching view of the work, and thus perspective on one's own contribution to the whole. From an author's perspective, the desire to be autonomous must be balanced against the desire to contribute to the overall document. "If soon after you get started," an author reflected on the need for early document leadership, "you can't get information from the project manager about what's in the other chapters (particularly if you're dependent on what's happening in other chapters), then everything falls apart."

For large documents in particular, document leaders fill the role of communicator. The experience related by one author emphasizes the need for a clearinghouse of information about goals and content. On the project he spoke of,

> ... the project manager was also lead author and that person pretty much assigned themselves the task of reading each chapter as it came in. So, it was kind of an editorial function almost, that they assigned themselves. Any contacts with the sponsor were through the project manager. So, you asked the project manager for information and the request was attended to by the project manager. So, the project manager knew what requests were out and what information had come back. That seemed to work fairly well. . . . You sometimes don't know what other people are doing who are writing at the same time you are. It is easy to fall into the problem of thinking, okay, this will probably be handled in another chapter . . . [or] adding into my chapter things that properly belong in another chapter. . . .

Such working knowledge is even more critical in coordinated documents. Because coauthors may work for corporations, government agencies, or universities that are widely scattered, there is a pressing need for consensus about the features and use of the written product. With a lead author, contributors assume that one mind has a sense of the whole; however, organizational constraints make it impractical to provide a lead author for each large project. This leads to varying degrees of consensus. Coauthors repeatedly express a need for consensus about the division of responsibilities in writing the drafts, in the content and scope of work expected (including organization), and in the uses of the written product. In this need can be seen a need for involvement in the "construction of [a] text world" because it incorporates coauthors' need for knowledge about the rhetorical situation, as well as the planning of their individual contributions (following Gunnarsson [15, p. 98]). Coauthors feel a need to make the process as well as the product.

DETERMINATION OF GOALS AND CONTENT

Those authors interviewed identified five influences on the goals and the content of their reporting. Two are common to all writers in the laboratory, whether individual or group: sponsors and the required document cycling process. The other influences are unique to multi-authored situations: lead authors/coordinators, proximity of the authors to each other, and the organizational norms among the authors.

One of the first goals of authors in this environment is to feel assured of their sponsor's intentions. One writer remembered an instance in which a new technical representative was appointed by the sponsoring organization. He engaged the laboratory to write a report discussing state-of-the-art spectroscopy and recommending new equipment for various uses. Because the technical representative was unavailable during the early stages of the writing, the authors grew uncertain of their sponsor's use of the report. In the process of reaching for a form for the report, they were stalled: "I think the biggest problem was that we were trying to shoot at a moving target. We did not sit down and nail down with the sponsors . . . exactly where we were going. So, I would write something and then we'd say, no, I can't really take this tack. . . . So, then, we'd have to sit down and think about it and rewrite it, and then we'd decide, no, that's not really the way we should go on it, either." In lieu of clear direction, the authors decided to set tentative goals, based on their hunches about the uses to which the sponsor would put the report.

Lead authors and coordinators can also become purveyors of goals and lodestones of content. The primary author of a manual of procedures described his lead authorship as an exercise in communicating goals and the document's readership and in reviewing his coauthor's drafts. First, he provided an outline and then discussed the report's purpose:

> . . . whether this document was going to teach or whether it would draw any conclusions. The reader might not know much about the technology, for instance, or the regulations that are involved with the technology. . . . I also asked for a copy of what [my coauthor] had got, even though it's not in a final form that he wants to turn in. He might not even call it a draft. . . . Then I'll look at it just to see whether he's covering the types of things I want covered and that he hasn't gone into too [much] detail about things that aren't important to the document.

A third influence on goals and content is the organizational review process. Writers interviewed rarely mentioned having achieved project goals as gauges of accomplishment. Instead, they cited a document's clearance of organizational hurdles. A common response to the question "How do you know when you 'have it right?'" was "when it's through clearance" or "when everything makes sense and looks logical and everyone else passes it through and says, yes, that looks good." The organizational procedures for authoring and reviewing guarantee that even a document with a single author can benefit from many reviewers evaluating the work reported, the format used, and the conclusions and recommendations generated.

A fourth influence on authors is proximity to their coauthors. When coauthors are spread far apart, the problems may become intensified. A document that was assigned to eight authors with seven affiliations at seven different addresses was divided into sections, with the coordinating author writing the introduction and summary. The group was an expert group, assembled by the sponsor by telephone and letter; sections were assigned by area of expertise.

After being assembled, the document was sent to the sponsoring government agency, where it was reviewed by several people and returned to the coordinator, who incorporated the sponsor's comments and sent the revised draft to each author for further revisions. One of the authors, reviewing the entire document for the first time, made major cuts in several chapters, on the assumption that the document was meant to provide procedures. His revisions were in keeping with his own contribution. That is, he revised the rest of the document to conform to his vision of the goals, perhaps formed early in the process of planning and writing. Most of the changes were rejected by the coordinator, who knew that the document was to be advisory and to include a discussion of past practices, both successful and unsuccessful. Although the overall purpose and readership were explained in a letter sent to each author at the beginning of the project, the deadlines and budget had not permitted the group to meet to discuss the technical or rhetorical elements of the work or the scope and progress of individual sections. Assigned their sections, the authors had apparently not considered it necessary to consult with one another and, as a result, formed idiosyncratic impressions of the document's goals and content.

The fifth influence on text construction is the social norms among the authors themselves. As the authors of documents are more comfortable with their coauthors and share more interpersonal interactions, there is more interaction on content. Two writers interviewed are parts of small working groups (four to seven scientists), who coauthor documents frequently and whose offices are close together. They maintain active and friendly relationships, both professionally and personally. These writers spoke of writing "strawmen" drafts in the initial stages of a document's development, much like the "shared document collaboration" noted by Allen et al. [14, pp. 84-85]. The "strawman" refers to an incomplete draft although different writers used the term for different degrees of completion: in a prewriting stage of brainstorming ideas, in the outline stage, or mostly complete. In one group, "strawman" referred to a role played by one of the authors as he or she presented the rough cut of a report or a set of rough ideas to the group. Such a "strawman" role was mutually acknowledged to advance the development of the document. The "strawman" understood that he was not presenting a draft for acceptance or rejection but for revision. Although the "strawman" often advocated his position, the rough nature of the draft encouraged the other authors to offer alternatives. In this situation, the draft was intentionally sacrificial, allowing a consensus to be built around rewording, reanalyzing the rhetorical situation or the data, and perhaps reconceiving the product to be presented.

RESOLUTION OF CONFLICTS

Doheny-Farina has shown dramatically that the writing environment not only influences the writing process, but that the writing process can also influence the environment [16]. In the case he describes, the writing process brought up conflicts that were long dormant in the company—the vice presidents were not pleased with the autocracy of the company's president and founder. The conflicts arose when they tried to offer an alternative to the business plan he had proposed, objecting that his plan was technically flawed and much too ambitious for their level of staffing and current knowledge. It was not that the writing environment in his example created the constraints and conflicts, but that it revealed them. In the company which he described, the written product was special, a vehicle to gain vital capital to continue funding their work.

The writing environment among scientists in a laboratory is rarely so dramatic as that described by Doheny-Farina though it is just as important to the future of the laboratory. With rare exceptions, the conflicts are more subtle and anchored in the content of the work being done: differing opinions about the most "elegant" (that is, simplest and most revealing) methods of experimentation or about conclusions to be drawn from the work. Writing in the highly structured environment of this laboratory offers the possibility of appealing to disinterested persons in the organization to resolve conflicts.

Disagreements about content are often referred informally to peers who are not involved in a multi-authored work as well as to one or more coauthors. "When I want to test myself, I go down the hall," said one of the scientists interviewed, "and talk to someone I find it easy to talk to." The practice of "testing" yourself to see if you're thinking rightly is common, particularly among areas in the department with an informal atmosphere. Such areas are evident by the open doors to offices and the writing-related discussions in the halls, lunch rooms, and other common space. As might be expected from communities of professionals, most substantive discussions are reserved for other professionals who would be familiar with the terminology and literature of one's work or with the sponsor for whom the work is being done.

Where the group norms are more formal or scientists prefer to work in private, other means are available for authors to "test" their ideas and obtain critiques of work. The clearance process requires a formal peer review, an editorial review, and three management levels of reviewers. Such scrutiny influences both the substance and the style of coauthored works. For example, a conflict over a conclusion in one coauthored work, a scientist reported, could not be resolved among the authors. There were good scientific reasons both for including the conclusion and for questioning its validity. The project manager, on the advice of two of the four coauthors, included the conclusion as a warranted interpretation of the data. The disagreement was eventually resolved by a higher level manager in the clearance process, who included the offending interpretation but only as a "possible conclusion" that had to be confirmed by more research. Overall, those interviewed placed a strong belief in the resolving power of the cycling process.

Less Solomon-like is the role of editors, who resolve conflicts over usage, format, and appropriateness for intended audiences. Indeed, the editorial role in resolving conflict among coauthors deserves more study. When editors are sure to review a document, particularly one that has lacked a strong central vision among the authors, some author groups rely on the editor to provide a stylistic coherence that a lead author or coordinator might not provide. The experience of the department's editors suggests that as a way of relieving authors' uncertainties about a work's homogeneity, authors may use editors to avoid potential conflicts over issues of content or usefulness to an intended audience. Paradis, Dobrin, and Miller found that the major tension in the document cycling process at Exxon ITD arose from supervisors editing writers' work, criticizing the writer's organization or reflection of purpose and audience [12, p. 300]. In the Battelle laboratory studied here, managers and groups of authors regularly turn the work over to an editor, rather than being forced to single out individuals for their verbal inadequacies. A good editor, those interviewed noted, can be valuable in resolving conflict by reworking a document's language around the technical content.

Wording, while apparently a small matter, may consume a considerable amount of effort among coauthors and be a source of disagreement. However, personal preferences in wording are apparently insignificant when compared to differences stemming from basic conceptual differences. In particular, coauthors can hold different views about the nature of language in reporting technical information. Those interviewed all remembered situations in which an author insisted on a wording that others thought sent the wrong message. Arguing over wording, nearly all reported, is common with some people. "There's one guy," a scientist related in informal conversation, "that just doesn't like his stuff changed. And he's real hard to work with in the last parts of a project." Some writers report feeling a disturbing disjunction between the technical matter and the linguistic matter. The agreements experienced by coauthors in planning and analysis stages are compromised when the whole work must be characterized for readers who may know nothing about the work's background and contribution.

It could be that conflicts arise between those authors who see language as a relatively rigid construct, with few truly synonymous expressions for the phenomena they describe, and those who see language as a more flexible instrument, with options (however constrained) for speaking of many technical phenomena in more than one way. Coauthors with a less flexible view of language may be loath to rephrase or reform information. In part, this betokens more an attitude toward language rather than any developed philosophy: an approach that recognizes language's ability to mislead, that sees language as a dearth rather than a potential. "If sometimes we could communicate a lot of this stuff without having to mess with words," a technician said, "it'd be a lot easier." However, there is also a sense that the technical consensus among authors, clear and unrefutable as the data may be, becomes fragile when it is bombarded with words. Although the scientist's attempt to retain the meaning in its technical sense has traditionally involved striving for an "objective" language, in fact, what may sound objective to a writer could appear to be a series of idiosyncratic choices to someone else, such as a coauthor.

CONSTRUCTION OF A COMMON VISION

Authors who work together in this environment need minimally three sources of knowledge to reduce the uncertainties and create a common working vision of the task.

1. Leadership: Authors need leadership—whether a lead author or a document coordinator—to keep them informed about the nature and extent of the contributions expected of them.
2. Goals/Content: Authors need clarity about use of the final product (to satisfy sponsors' needs); such information coheres the contributions to the final product by providing a basis for choices about relevant information and casting of that information.
3. Conflict/Resolution: Authors need social/organizational review to resolve disagreements and negotiate consensus; this may require a flexible view of language to overcome the authors' resistance to having their work changed by others.

By providing these kinds of knowledge, an organization can ease the burdens of working together and increase the opportunities for success. In large projects, it is vital to overcome the alienation of authors by increasing the extent to which they share the common purposes of the document, understand their audiences' interests and uses for the work, and recognize the scope of their contributions. They need information about the social and organizational contexts which must approve the work. That is, each contributor needs a framework in which to see the task. The following recommendations are directed at aiding writers in developing their potential for consensus on a project:

1. Project or task managers should provide early and frequent information to coauthors. Assign a document leader who is involved in the writing of the document or who can answer questions about contributions and expectations.
2. For authors who are widely separated physically, at least one face-to-face meeting is preferable; however, other means may be available to ensure that authors know deadlines, formats, and rhetorical considerations.
3. Authors should have a vision of the whole in order to see how their work fits. Provide models or clear statements of the use of documents, as well as opportunities to test them. Make sure the nature of the end product is commonly understood.
4. Project or task managers should assist coauthors in determining how to best review their work informally, to create a collaborative sense.

REFERENCES

1. L. Flower and J. Hayes, Images, Plans, and Prose: The Representation of Meaning in Writing, *Written Communication, 1*:1, pp. 120–160, 1984.
2. T. Kuhn, *The Structure of Scientific Revolutions,* 2nd Edition, University of Chicago Press, Chicago, 1970.
3. C. Bazerman, Scientific Writing as a Social Act: A Review of the Literature of the Sociology of Science, in *New Essays in Technical and Scientific Communication: Research, Theory, and Practice,* P. Anderson et al. (eds.), Baywood Publishing Company, Amityville, New York, pp. 156–183, 1983.
4. K. Bruffee, Social Construction, Language, and the Authority of Knowledge: A Bibliographical Essay, *College English, 48*:8, pp. 775–790, 1986.
5. B. Latour and S. Woolgar, *Laboratory Life: The Social Construction of Scientific Facts,* Sage Publishing, Beverly Hills, California, 1979.
6. L. Odell, D. Goswami, and A. Herrington, The Discourse-Based Interview: A Procedure for Exploring the Tacit Knowledge of Writers in Nonacademic Settings, in *Research on Writing: Principles and Methods,* P. Mosenthal, L. Tamor, and S. Walmsley (eds.), Longman Publishing, New York, pp. 221–236, 1983.
7. L. Odell and D. Goswami, Writing in a Non-Academic Setting, *Research in the Teaching of English, 16*:3, pp. 201–223, 1982.
8. J. Fahnestock, Accommodating Science: The Rhetorical Life of Scientific Facts, *Written Communication, 3*:3, pp. 275–296, 1986.
9. G. Myers, The Social Construction of Two Biologists' Proposals, *Written Communication, 2*:3, pp. 219–245, 1985.
10. C. Argyris and D. Schon, *Theory in Practice: Increasing Professional Effectiveness,* Jossey-Bass Publishing, San Francisco, California, 1974.
11. M. LaRoche and S. Pearson, Rhetoric and Rational Enterprises, *Written Communication, 2*:3, pp. 246–268, 1985.
12. J. Paradis, D. Dobrin, and R. Miller, Writing at Exxon ITD: Notes on the Writing Environment of an R&D Organization, in *Writing in Nonacademic Settings,* L. Odell and D. Goswami (eds.), The Guilford Press, New York, pp. 281–307, 1985.
13. H. Wiener, Collaborative Learning in the Classroom: A Guide to Evaluation, *College English, 48*:1, pp. 52–61, 1986.
14. N. Allen, D. Atkinson, M. Morgan, T. Moore, and C. Snow, What Experienced Collaborators Say About Collaborative Writing, *Journal of Business and Technical Communication, 1*:2, pp. 70–90, 1987.
15. B. Gunnarsson, Text Comprehensibility and the Writing Process, *Written Communication, 6*:1, pp. 86–107, 1989.
16. S. Doheny-Farina, Writing in an Emerging Organization: An Ethnographic Study, *Written Communication, 3*:2, pp. 158–185, 1986.

CHAPTER 4

Insight and Collaborative Writing

MEG MORGAN
MARY MURRAY

Given a choice, many people prefer not to work in groups. Group work takes a lot of time and energy; it requires interpersonal skills, tolerance for others' opinions, discrimination of the ridiculous from the merely absurd. It is often less efficient and more cumbersome than working alone. Group, or collaborative, writing (two or more people working together on a single document) shares the same costs with all group work—it is time-consuming, it requires an openness not only to others' opinions but to others' writing styles. It requires that collaborative writers "check [their] ego[s] at the door" as they negotiate everything from word choice, to organization, to whose name appears first on the title page [1, p. 83]. Yet, despite its costs, collaborative writing is used in nonacademic settings and has been the focus of increasing interest among writing researchers over the last few years, including studies in the Odell and Goswami collection [2] and Doheny-Farina's ethnographic study of writing in an emerging software company [3]. But why does collaborative writing exist at all considering its costs? The answer is clear: Groups, for all their problems, under certain circumstances and with certain tasks, produce better solutions to problems than individuals working alone [4].

In this chapter, we will look at one aspect of the phrase, "better solution;" we will prepare readers to recognize an insight by looking at insight through the eyes of scholars in several disciplines. We will discuss the nature of individual insight and the conditions under which an individual insight might occur. We will then apply these notions of individual insight to small problem-solving groups, examining a collaborative writing group to answer some questions about insight. Finally, we will look at a case of non-insight—a situation which meets the conditions for insight but one in which an insight did not occur, and, using small group research findings, we will speculate on the reasons for the failure of the group to have an insight.

We suspect groups are as capable as individuals of achieving insights because, as you will see, some of the conditions for insight which must be *stimulated* in an individual are often *givens* in a group situation. Yet the study of insight, in groups or individuals, especially the empirical study of insight, is similar to the study of hurricanes. You can do all the preparatory research you want, studying wind, temperature, water, the jet stream, droughts in West Africa, etc. You can anticipate a hurricane because the right conditions exist. But you cannot force one to occur no matter how interested you might be in the hurricane phenomenon or how promising the conditions to produce one. But if one does occur you will recognize it when you see it, will be able to predict its behavior, and will know its effects.

WHAT IS AN INSIGHT?

Shea, a theologian, tells the insight story of a retired couple who began flying their grandchildren down to their home in Florida, simply to get to know them. "You've got to get to know your grandchildren," said the man to a stranger on a business flight [5, p. 23]. A student in Professor Murray's class wrote about a meaningful place, the gravesite of his grandfather. Franklin and his black family had moved north from Arkansas leaving the grandfather behind. Franklin wrote of a boring July 4th when he travelled back to Arkansas and he and his family decided to find the grandfather's grave. The essay details a brown, overgrown cemetery covered by lush green vegetation, and a long search to find the grave amidst the underbrush. Franklin's insight is carried in the image of the essay's final paragraph: the family silhouetted against the July 4th evening sky. Nature seems to sanction both the freedom chosen by the family's northern migration and the respect shown the grandfather by the family's search for the grave.

Both these stories contain classic elements of what scholars in a variety of disciplines term insight. The presence of an existential need, confrontation, the use of the intellect and of the emotions—all prefigure insight. The old couple chose the cost of flying their grandchildren over isolation. Franklin felt uneasy about the gravesite and wrote about his experience to make sense of his feelings.

The effects of insight are action, peace, and the ability to use the insight on future occasions. We've all been charmed by the maxims of strangers and friends alike, which are really distilled insight experiences. When the grandfather uttered his maxim to the stranger, he was revealing a truth that he knew in a very personal way; he is likely to use that knowledge in future situations as well. Both the couple and Franklin feel a peace that comes from confronting and resolving something both puzzling and meaningful.

Insight itself is a radically new way of seeing the subject that discloses intellectual boundaries and is simple yet permanently true. Perhaps the couple struggled over money; yet the insight revealed to them (and to us) that life (their own lives and life in general) is more than material things. Franklin may have felt guilty for leaving his grandfather; the green gravesite, however, taught him of the hope and life that his grandfather held out to his family. Franklin and his family walked away in peace, knowing that the decision for freedom and a better way of life was the right one (ironically rediscovered on the 4th of July). So, what is insight? Insight is: a solution that completely resolves a multi-faceted dissonance that has been confronted emotionally as well as intellectually. An insight seems to come suddenly and without effort. It is a new understanding, according to Elbow, a "shape where a moment ago there was none" [6, p. 35]; theologian Baum considers it a new awareness [7]; another theologian, Dulles, "a breakthrough to a higher level of consciousness" [8, p. 115].

HOW IS INSIGHT STUDIED?

The empirical study of insight is difficult because insights, like hurricanes, are not amenable to summons. Even if the conditions are right, an insight may not occur. Although difficult to observe in an empirical setting, it is still possible to understand insight, especially by taking an interdisciplinary approach. The disciplines that inform our examination of human insight are philosophy, education, theology, composition, and psychology. One would think that studying psychology alone would produce reliable information about insight; however, despite their constant use of the term, psychology researchers seldom agree on its meaning and have conducted few large-scale studies of the construct. Therefore, it is necessary to gather the qualities of insight from a variety of fields in order to achieve a reliable understanding of the term.

Readers may find our references to these fields odd. Yet experts' accord on the nature of insight—despite the diversity of methods and goals for studying it— lends validity to our approach and our findings. Theologians, such as Baum [7], Dulles [8], Fawcett [9], and Shea [5], study insight to understand the nature of divine revelation in the individual; philosophers, such as Lonergan [10], study insight as a type of human knowledge; composition scholars, such as Young, Becker and Pike [11], Elbow [6], and Berthoff [12] investigate insight as a function of writing; education experts, such as Boyer [13] and Sloan [14], examine the role of imagination and insight in the learning process. Finally, psychologists, such as Piaget [15], Festinger [16], and Bruner [17], examine learning behaviors which include inquiry and insight.

In the next section of this chapter, we will discuss the three conditions necessary to produce an insight: dissonance, confrontation, and the involvement of the whole person.

WHAT ARE THE CONDITIONS OF INSIGHT?

All the scholars mentioned above agree on three conditions for insight: dissonance, confrontation, and whole-person involvement. By dissonance we mean a feeling of incongruity, puzzlement, wonder. Hence, dissonance can be positive or negative: positive in the happy wondering of why a new home far exceeds expectations or negative in the puzzlement of why a new product is not accepted by a predicted market. In the rest of this chapter we will configure dissonance as a problem because most collaborative groups exist to solve problems, not to explore happy wonderings. In its broadest sense, however, dissonance certainly includes positive puzzlement. To summarize, an insight is more likely to occur when

- a multi-faceted problem is recognized;
- the problem is confronted in its fullness and complexity; and
- the problem is considered not only with the intellect but also with emotions, attitudes, and beliefs.

Dissonance

Unless a person internalizes a problem, insight is not likely to occur because insight is, according to Lonergan, "a function, not of outside forces, but of inner conditions" [10, p. 5]. Internal recognition of a problem, referred to as "dissonance" by psychologist Festinger, is a condition of incongruity, puzzlement, wonder, or curiosity [16]. This dissonance that takes place at the beginning of the learning or problem-solving process increases our desire to explore [15-17]. In other words, when a person cannot relate new knowledge to previous knowledge, she may become confused and begin to inquire. In education studies, the lack of insight has been connected to the lack of imaginative learning. According to Boyer, schools have become cafeterias of information rather than places where students puzzle over new relationships, encouraged by teachers to see how disciplines and theories intersect [13].

The importance of dissonance also permeates the literature of theology. The theologians mentioned above have been studying the nature of individual revelation or insight; Fawcett places the "presence of existential need" or dissonance at the beginning of the faith experience, and Dulles also claims that dissonance precedes insight [8, 9]. Theologian Shea makes the connection between internalized dissonance and insight explicit:

> The needs that are troubling us, the drives that are urging us on, the conflicts that we are engaged in shape the content of the revelation. This does not mean that revelation is caused by the needs, drives, or conflicts but that they make us receptive, gear us to specific communication [5, pp. 24-25].

Composition experts both explicitly and implicitly echo what scholars in other disciplines believe: until there is dissonance, there is little chance for insight. Young, Becker and Pike argue that only out of incongruity can insight emerge [11]; Berthoff states that the chaos writers feel when puzzled over either what to write or how to write it can generate insightful material [12]. When writing is seen as a problem-solving process, *a la* Flower and Hayes, the critical first step is problem recognition, a configuration similar to incongruity and chaos [18].

Confrontation of Dissonance

Confronting dissonance means accepting a problem on its own terms, accepting all facets of the problem, not confusing the problem with similar problems or reducing it to a simpler form. It means exploring the problem fully, considering its full context and complexity; confronting dissonance, according to Paul Tillich, is "always correlated with human questions arising out of a specific cultural and historical context" [19, p. 102]. Thus the problem solver must recognize the uniqueness of a particular problem and be willing to face it. The person who refuses to confront dissonance will never achieve an insight, according to linguist Labov and psychologist Fanshell [20].

On the other hand, Lonergan notes that a person may confront the dissonance, but flee from the insight. According to Lonergan, fleeing from insight "generates misunderstanding both in ourselves and in others;" allowing fantasies or old ways of understanding to overrule the possibility of attaining new insights risks sacrificing human development [10, p. 191]. He states also that the flight from insight can lead to human isolation.

Fleeing from insight also can take another form: reformulating the problem into a more familiar shape so the refashioned problem and its solution become more familiar. For example, an organization may have a problem retaining employees in a certain department. Instead of addressing the difficult and perhaps unfamiliar issues inherent in management style, benefits, workload, social milieu, and interpersonal relations which may be related to lack of employee retention, the organization considers the problem an environmental one, and buys everyone new desks and potted plants. The organization reformulates the problem to make it more familiar, perhaps similar to ones the organization has previously solved. However, the organization's unwillingness or inability to face the whole problem makes the "solution" temporary at best.

Confronting dissonance is always a matter of confronting the unknown, an agreement to move from the familiar to the unfamiliar, from the comfortable to the strange.

Whole-Person Involvement

The involvement of the whole person is the third condition for achieving insight. Solving problems usually requires intellectual efforts, such as calling in an expert or researching various options. But when insight and not simply an intellectual solution is the goal, the whole person must be involved: the intellect, the emotions, the behavior, and the attitudes. A diversity of persons in a collaborative writing group (women, minorities, people with different training and abilities) can be a positive source for generating insight because they provide a rich display of emotions, attitudes and intellectual approaches.

One key to whole-person involvement is to include the emotions and attitudes in the questions asked to confront dissonance. Compositionists Lauer, Montague, Lunsford, and Emig show writers how to ask questions that pull together the subject of the inquiry (thought) with the writer's attitude towards the subject (feelings) [21]. This bi-modal question-asking strategy encourages a whole-person response to the dissonance.

Psychologists Klein, Mathieu, Gendlin, and Kiesler make the integration of emotion and cognition their definition of insight. They designed an Experiencing Scale, the highest level of which describes the patient as able to give a "full, easy presentation of experiencing" where "all elements are confidently integrated" [22, p. 64]. Psychotherapists Strupp and Binder consider insight the "affective experiencing and cognitive understanding of current maladaptive patterns of behavior that repeat childhood patterns of interpersonal conflict" [23, pp. 24-25].

That problems require the confrontation from the whole person can be seen from writers in a variety of fields. If insight is defined as a "new awareness" or "breakthrough to a higher consciousness" as theologians Baum and Dulles claim it is, how can the intellect alone transport thinking to that level? It cannot. Shengold, a psychologist, claims that only thinking that enables us to step out of a routine way of envisioning a problem yields insight [24]. And philosopher Fontinell and theologian Rahner agree that insight is noncognitive [25, 26]. Fontinell links it to symbolic thinking which may be intellectual, but a fast and seemingly intuitive approach. Rahner calls it "a state of mind—not knowledge but a consciousness" [26, p. 411].

The condition of whole-person response has an interesting twist in the area of gender studies. Researchers Belenky, Clinchy, Goldberger and Tarule attest to women's ability to learn in noncognitive ways [27]. In their interviews of 135 women concerning learning styles, a large percentage reported that experience and intuition were their ways of knowing, not the intellectual approach of the academic setting in which they found themselves. Many communities, such as the academic community, may not draw on women's experience and intuition as "information" (which women reported as an alienating experience). Their way of approaching problems yields a new perspective that may foster insight.

INTEGRATING QUESTIONS

Assuming that a group can experience dissonance, confrontation, and whole-person response (and we will explore this assumption later in this chapter), these questions arise:

1. In what ways can a collaborative writing group experience dissonance? What are some "symptoms" to indicate that a group is in a state of dissonance?
2. How does a collaborative writing group confront its dissonance? How are interpersonal conflict in a group and confrontation of dissonance related?
3. How can a group respond both cognitively and affectively to dissonance?
4. How can the conditions for insight be present and an insight still not occur? What factors seem to obstruct achieving insight in a collaborative writing group?

We will try to answer these questions in the rest of this chapter.

INSIGHT IN COLLABORATIVE WRITING GROUPS

The investigation into the possibility of insight in small groups creates intellectual tensions between theoretical know-how and common sense. Given the conditions of insight, dissonance, confrontation, and whole-person response, the group—and its individual members—can certainly experience dissonance and confrontation. In fact, a group is often assembled to help resolve uncertainty because groups are superior to individuals in those tasks where individuals are likely to have an insight—complex tasks, relatively ill-defined or abstract problems [4]. Talking to other group members—something interacting groups must do that individuals may not be able to do—is also beneficial to decision-making when groups work on complicated tasks, those that involve values, or in situations in which no group members possess especially high levels of relevant knowledge, the very tasks where an insight may be most needed [28-30]. In addition, interacting groups have also been shown to produce higher quality decisions on a complex information evaluation task [31]. Thus, both groups and individuals are predisposed by the task to share two conditions for insight, and these are the very conditions under which groups outperform individuals.

However, another tension exists: although it is possible for an individual to have a whole-person response to a dissonance, is it possible for a group to act as a single unit, a group-as-one? Small group research evidence suggests that such a phenomenon may indeed exist. Cattell developed a theory of small groups, group syntality theory, which suggests that a group has a distinctive personality, can "take on an identity, and that it should be viewed as an organic, functionally integrated entity" [32, 33, p. 114]. The "assembly effect" approach also suggests that a group can act as one. According to Collins and Guetzkow, "[t] he assembly effect bonus is productivity which exceeds the potential of the most capable member and also exceeds the sum of the efforts of the group members working separately" [34, p. 58]. Both Cattell and Collins and Guetzkow suggest that groups establish unique identities, which suggests that groups may respond as single units, both cognitively and affectively, in ways that can produce insight.

Oddly enough, although the individual has been the focus of much of the research into insight, insight in groups is in fact more accessible to researchers than insight in individuals. An individual conducts much of the activity necessary for insight through internal conversation—talking to him/herself, thinking, taking notes, or making journal entries. A group will not have an insight unless it vocalizes its dissonance, confrontation, and whole-group response; and once the group vocalizes, feelings, thoughts, and questions, often buried within an individual, are accessible to group members and others—including researchers.

It is important to notice the emphasis on "group interaction" in the previous discussion. We believe that conversation in groups creates the dissonance (or type of dissonance), enables the group to confront the dissonance, and establishes the group identity. Group communication, of course, does more; according to Poole, it creates factors such as roles, norms, group structures, and a group climate [35]. In fact, communication within the group constitutes that group's reality. Although realities exist outside the group (such as personality traits or leadership characteristics), these realities must be enacted in the group if they are to affect the group's behavior. Putnam maintains that patterns of conversation or interaction "evolve over time and through sequences of messages. Thus, ... the behavior of one member alters, intensifies, or inhibits the actions of another" [36, p. 186]. This theory of group interaction, where communication in a group constitutes the group's reality, is called the theory of structuration; adopting the theory of structuration as a theory for small group interaction eliminates the view that behaviors within a group are controlled by factors outside the group, a view Poole calls "deterministic." In terms of insight, a group will have an insight based not on variables that can be measured independent of the group (such as leadership, I.Q. of members, or group size), but on the patterns of interactions that occur within the group.

To return to our original point: accessibility. Because we make interaction or conversation central to understanding group behavior, and because conversation is so accessible, insight in a small group is also accessible. Although an insight may appear to come suddenly, it is actually the result of considerable work, apparent in the group through its conversation. But because insight itself cannot be managed or arranged, watching for group insight has the same accidental quality as watching for individual insight.

Let's discuss the three conditions of insight in the context of small problem-solving groups. In small groups that meet to solve a problem, dissonance is associated with the problem. A group meets to study ways to improve staff morale; after it studies the problem it makes recommendations to the supervisor who may accept or reject them. The group experiences dissonance, although complex, only in relation to the morale problem.

Collaborative writing groups are organized to solve a problem but they are also charged to frame their solution in the form of a written document—a book, an article, a proposal, a position paper [1]. As such, they must solve two complex problems: the organization's problem and the writing or task problem. Both kinds of problems involve dissonance; both kinds of problems may benefit from an insight. And with both kinds of problems, to achieve an insight, the dissonance must be internalized. When either an organization's problem or a writing problem is *handed to* a group (or individual, for that matter), it may not become internalized. Even if one member of the group has internalized the problem, it does not follow that the whole group (not individual members but the group as a whole) accepts the problem. Whole-group recognition and acceptance of a problem is one of the critical conditions for group insight and may be observed through group communication. For example, a group may agree that no one solution will satisfy everyone in the organization; thus, the group decides to find the best solution, even though members know some people will be hurt or offended.

Although some problems are given to groups to solve, others emerge in a group in the process of its work. Often a group redefines a problem in the process of conversation. What begins as a software user problem becomes an employee competence problem, and dissonance rises as members of the group struggle to negotiate meanings held by individual members and shape them into a shared meaning. In addition, as the writing task proceeds, the group may experience dissonance over the constraints of a writing task. For example, a writing group may struggle to understand the needs of a particular audience for its written document.

Thus a collaborative writing group must negotiate two kinds of problems and the dissonances that may accompany them: the organization's problem and the writing task problem. This dual dissonance makes the work of the collaborative writing group complex and difficult.

The second condition for insight is confrontation. In a small group, as in an individual, confronting a problem is recognizing the uniqueness of the problem, how it differs from other, perhaps similar, problems. A group that has confronted a problem can ask a specific question about that problem in order to try to solve it. Confronting a problem may manifest itself in a group through conversations among members that try to pare away interfering details to get at the heart of the matter. Confrontation also presumes a full exploration of the problem in all its complexity, openness to questioning, and acceptance of the members' understanding of the problem.

The third condition of insight, whole-person involvement, manifests itself in small groups through conversations that are noticeably both cognitive and emotional. For example, the group discusses changes in child abuse legislation, considering child abuse a product of social attitudes toward children as property (cognitive) as well as a horror against the innocent and helpless (affective). In the group conversation, both aspects of child abuse will emerge. Cognitive and affective involvement may be possible in even less volatile issues. A recent discussion of curriculum changes in a college academic department produced conversations that centered both on the need to recognize employment patterns in today's business market (cognitive) as well as the morality of educating students, some of whom can only find jobs paying just above the minimum wage (affective).

Both confronting the problem and responding to it cognitively and affectively are necessary conditions whether the dissonance is associated with the organization's problem or the writing task problem.

SOME EMPIRICAL EVIDENCE

We would now like to present some empirical evidence to suggest that groups, like individuals, can experience the conditions of insight. However, even given the presence of these conditions, an insight may not occur. Our example shows that if a group had arrived at an insight it could have solved all facets of its complex problem. However, certain conditions existed which seemed to prevent this group from coming to insight, conditions which reflect some of the conditions presented in this article.

In fall 1986, Professor Morgan videotaped four collaborative writing groups, each in the process of planning and drafting a research proposal as an assignment in a business writing class. The taping was part of a major research project to examine collaborative composing processes, including how writing groups define problems, how they divide the writing task, what behaviors they exhibit during drafting, and how leaders emerge in the groups. Examining insight was not one of the intended purposes of the research study.

Students met in four different groups, each for three 50-minute classes. The students were placed in their respective groups by their teacher, who, on a random basis, assigned students with the same grade in the course to the same group. The maximum number in each group was four. One group was made up of students with "A" averages; two groups were made up of students averaging "B;" and one group was comprised of students with "C" averages. The writing assignment for all groups, including those not selected for taping, was to identify and explore a problem in an organization, to draft a proposal to a decision-maker in that organization outlining research methods the groups will use to find solutions, and to revise the draft of the proposal before submission. (The proposal is the first document in a three-document, six-week project.) Approximately three weeks after students hand in the revised proposal, they will group write and revise a progress report; in another three weeks, they will group write and revise a long, analytical report. For a more complete description of this assignment, see Morgan et al. [37].

Three sessions were videotaped; at the end of the third session, students were to have decided which organizational problem they would explore and were to have written a first draft of the research proposal. No directions were given to the students other than a first draft was due at the end of the third session.

The group we examined for insight in this article was the "A" group, a group made up of three female students (Dianne, Jackie, and Patti) and one male student (Robin). Dianne and Jackie knew each other before working in this group because they belonged to the same sorority.

Professor Morgan was not looking for insight in these groups; however, during her observation of Group "A" an occasion arose in which the group experienced considerable dissonance. The dissonance arose not in connection with the problem for which the group was trying to find a solution (group members had resolved any dissonance here to their satisfaction at earlier meetings), but in connection with the writing task. Although an insight would have resolved all facets of the group's dissonance, an insight did not occur. Now we will examine the situation and communicative interactions that helped account for the lack of insight and try to explain this lack of insight using a framework from small group research.

At the first meeting, the four group members decided that they would investigate alternative ways to distribute basketball tickets to the university community—the organizational problem; no drafting occurred at this meeting. The members discussed their different perceptions of the problem, and, unable to define it to the group members' mutual satisfaction, they all agreed that they needed additional information and decided to talk to experts outside the group about ticket distribution at other universities before the next meeting. Jackie and Dianne established themselves at this meeting as the most talkative members of the group; their combined interactions accounted for 88 percent of the group's comments five seconds or longer.

At the second meeting, after sharing the information group members acquired between meetings about ticket distribution, and coming to a common understanding of the problem and possible solution, the group began to draft the research proposal. Only three members were present, Dianne and Jackie, the two sorority sisters, and Robin, the male student. They struggled to draft the proposal, writing one or two sentences but then crossing them out only to write one or two more. The group finally happened upon a technique that helped them draft: they began to imitate one of the samples in the course textbook. Jackie read aloud a sentence from a proposal in the textbook after which the other two members composed a sentence for their own proposal based on the one read to them. After they adopted this method, the draft proceeded much more smoothly and the group successfully completed the opening paragraph of the proposal. Group members decided to finish the proposal at the third meeting. Jackie and Dianne dominated this second meeting also: as Jackie read aloud from the textbook, Dianne generated 75 percent of the words written during this meeting.

At the third meeting, Robin was absent, although Patti was present. This new group continued drafting the proposal. However, during this meeting, the group experienced a great deal of writing-based dissonance. Dianne and Jackie, the dominant group members, confused the preliminary research they did between meetings with a full-blown investigation that normally follows the *acceptance* of a research proposal. Thus, they believed their research proposal was a recommendation report. Because they were confused, the strategy of generating sentences from the sample proposal they used during the second meeting did not work. For example, the textbook sample used the future tense ("I will interview . . .") to outline future research, whereas Dianne and Jackie insisted they had already conducted their research so the future tense "doesn't fit." At one point, Dianne asked: "This is the long report, right?" Patti (who never participated in the reading/writing strategy used at the second meeting) responded, "No, this isn't the long report."

The critical interaction among the group members during which an insight could have solved the group's dissonance occurred when Patti asked for clarification about the research methods from the sorority sisters; her questions suggest that she knew more research was necessary.

> Patti: Aren't we going to go out and talk to anybody between now [and the long report]?
> Jackie: We already talked to people.
> Patti: Oh, did we?
> Dianne: Yea.

Then Jackie enumerated the people they talked to between the first and second meetings.

> Dianne: So, since we already did all that . . . I suppose we could go . . .
> Jackie: I don't think we need to.

Patti's two questions ("Aren't we going to go out to talk to anybody?" and "Oh, did we?") plus her previous statement that the document was not the long report suggest she knew the group was on the wrong track. But Patti asked no more questions after this, and, in fact, helped draft the misguided proposal.

An analysis of the group's enactment of the three conditions for insight suggests that the group possessed the requisite dissonance. Dissonance is clearly evident in the group's frustration over not being able to use the matching strategy that worked at the previous meeting. This failure to fit together the sample in the textbook with members' concept of the document they were writing created confusion, puzzlement, and inaction. More important, however, is the group's failure to *confront the dissonance* which prevented group members from asking the right question to help resolve it: Is the research we performed between meetings the same as the research we must perform to solve the problem? Thus, it is the lack of confrontation rather than the lack of dissonance that keeps the group from the insight.

In order to see this more clearly it is necessary to expand our discussion of confrontation. In the case of individual insight, confrontation includes the willingness of an individual to acknowledge the dissonance, to see a problem as separate from other problems, and to ask the right questions to isolate the dissonance. Confrontation suggests a face-to-faceness with the unknown. In a small group, this confrontation may take at least two forms: 1) confrontation with the dissonance and 2) confrontation with other members of the group. In Group A, these two types of confrontations are bonded together. Because Patti was unwilling to confront other group members, the group did not productively confront its dissonance and see the problem from a unique stand-point; it fell back on a familiar pattern which, under new circumstances, did not work—imitating the sample from the textbook—and persisted in this pattern. Why didn't Patti state the need to gather additional data? Why didn't she challenge Dianne and Jackie?

Small group discussions of decision-making and conflict offer some answers to these questions. Hirokawa and Scheerhorn say that faulty decision-making can be the product of influence on the group by one or more members [38, p. 77]. In the case of Group A, the more vocal members of the group, Dianne and Jackie, seemed able to quiet the doubts of a less vocal group member who had missed one meeting.

Small group literature in the area of conflict helps answer these questions by suggesting three explanations for Patti's failure to assert herself and perhaps achieve insight:

1. the forming of coalitions and power relationships within the group;
2. the deviation from norms and roles established in the group; and
3. the failure of the group to experience important phases of development [36, p. 186].

The next section of this chapter will discuss each of these in turn.

COALITIONS, DEVIANCE, AND CONFLICT

According to Putnam, coalitions, "subgroupings of two or more people who unite to combine resources, provide mutual support, and influence decision outcomes", can control the direction of the group [36, p. 186]. The two sorority sisters had already formed a strong coalition based on both their previous association in the sorority and their presence at all three meetings. They dominated group discussion during planning, and contributed most content to the proposal during drafting. Patti seemed to accept her position as outsider, as she did not challenge the coalition by engaging in conflict.

Patti also exhibited deviant behavior which tended to strengthen the coalition of the other two members. She was deviant because she missed a meeting, departing from the norms or expectations of the other members who missed no meetings. She was also deviant because she questioned an idea—about the research that had yet to be performed. The two sorority sisters reinforced her deviant position by categorically rejecting her ideas.

Finally, Fisher's phases of group development suggest that conflict in a group occurs only after a period of group orientation [39]. The third meeting of this group was in fact the first meeting of an all-female sub-group. This group of three women was still in a period of orientation according to group development research. According to Fisher's model, the near-completion of a task would have already included a conflict phase in which group members aired differences and explored different viewpoints; this group, whose membership had changed three times on three occasions, managed to complete the draft of the proposal—its task—while arrested at the orientation phase of group development. Thus, the group never experienced a period of confrontation which might have made the proposal a successfully written document.

The lack of confrontation in this group, which stems from an incongruity between the progress toward completion of the task and lack of progress through Fisher's phases, seems to lead to the group's unwillingness to consider the problem in all its complexity. When Patti suggested that perhaps research had yet to be done, the coalition avoided the confrontation with dissonance that acknowledging Patti would require. Patti herself avoided the personal confrontation by remaining silent after a rebuff.

In addition, confronting dissonance in all its complexity assumes the problem-solvers know all the complexities of the problem. In this group, members were unprepared to write the research proposal. Not only had they never written one before, they had not read the textbook which may have prepared them. During the third meeting, all three women paged through the textbook for something to help them draft. Consequently, both Dianne and Jackie seemed to misunderstand the concept of research proposal and Patti's knowledge was so uncertain and her position in the group so tenuous that she did not attempt to clarify the situation.

The third condition of insight is involvement of the whole person. It follows that a group that refuses to confront a dissonance cannot be cognitively and emotionally involved in resolving it. During this third meeting, the group members were more concerned with producing a draft of the proposal than understanding the dissonance. For example, the group spent almost 10 minutes discussing the differences between "Procedures" or research methods and "Schedule" or day-by-day outline in their proposal. They were confused because the samples in the textbook clearly separated research methods from time schedule by putting them into separate segments. Because the group believed it already performed the necessary research, it could not generate a schedule for performing the research. As Jackie stated, "Our procedures are our schedule." At no point did any member say "This isn't right because what we're doing doesn't make sense;" nor did any member suggest discomfort with what was happening in the group. Both cognition and affect seemed to be ignored.

Consequently, the group wrote the research proposal as if it were a recommendation report, then submitted it to the instructor who subsequently informed the group in a conference that they could not recommend a solution because they had not done the research.

THE BENEFITS OF INSIGHT

The benefits of insight are clear, both for the individual and the group. First, insights lead to action. Without the insight that what they were writing was a proposal, that the research they performed earlier was related to defining the problem not producing a solution, Group A's members remained confused and puzzled, almost frozen. In fact, the group seemed to fragment, incapable of acting in concert. Dianne wrote the closing to the proposal almost singlehandedly, copying it almost word-for-word from the sample in the textbook. An insight, on the other hand, would have enabled the students to progress: to use the preliminary research as evidence of the problem, and to plan a strategy for finding a solution.

Second, insights bring relief or peace because the problem has been solved. Group A's members ended the third meeting quietly, by closing their textbooks and discussing who would type the proposal for the teacher conference. There was no discussion of a job well done, no words of congratulations among group members. If there was any relief, it was relief that a confusing meeting had ended with the writing task accomplished, successfully or unsuccessfully.

In addition, insights create new understandings—they change the way we look at things, at the world. The group's meeting ended with no new understandings. It was not until the teacher conference that the students seemed to understand the relationship between a report and a research proposal.

IMPLICATIONS FOR COLLABORATION

Because collaboration is used with increasing frequency in both the professional and academic worlds, it is important to learn some things from this chapter on insight and non-insight. First, collaborative writing groups acting as one can experience dissonance, but the dissonance seems more complex than that experienced by non-writing groups. Writing groups must engage in two complex tasks, problem solving and writing, both of which are capable of producing dissonance. A group in a state of dissonance is confused, puzzled and immobile. Members ask questions, but the answers do not satisfy.

Second, confrontation necessary to achieve insight in writing groups may take two forms: confrontation with the problem and confrontation among members. Productive confrontation requires that group members accept each other; it also requires, it seems, a level of group cohesiveness achieved only after a group has developed sufficiently. Consequently, confrontation of dissonance may not occur in newly formed groups where interpersonal conflicts are likely. Confronting the problem requires that a group be able to explore it fully; thus all information necessary to solve the problem with an insight has to be made available to the group.

Third, both cognitive and affective responses to a dissonance may be more possible if difference of opinion is allowed, if deviance is tolerated, and if the groups themselves include a diverse membership. The much feared "group think" phenomenon is not possible in a cohesive group comprised of members representing multiple perspectives. Certainly, insight seems more likely in that kind of group.

Organizations that value insights, new ways of looking at problems, must tolerate in its employee groups dissonance in all its complexity and confusion, must permit confrontation among group members, and must validate and welcome the emotions, intuitions, and cultural values of its members.

Our example of non-insight shows the confusion, waste, and diffusion that resulted from the students' failure to be prepared and open. We believe that group insight is possible if groups create and work within an environment that encourages questioning and acceptance.

REFERENCES

1. N. Allen, D. Atkinson, M. Morgan, T. Moore, and C. Snow, What Experienced Collaborators Say about Collaborative Writing, *Iowa State Journal of Business and Technical Communication, 1*:2, pp. 70-90, 1987.
2. L. Odell and D. Goswami, *Writing in Nonacademic Settings,* The Guilford Press, New York, 1985.
3. S. Doheny-Farina, Writing in an Emerging Organization: An Ethnographic Study, *Written Communication, 3*:2, pp. 158-184, 1986.
4. G. Hill, Group Versus Individual Performance: Are N+1 Heads Better than One? *Psychological Bulletin, 91*:3, pp. 517-539, 1982.
5. J. Shea, *Stories of Faith,* Thomas More Press, Chicago, Illinois, 1980.
6. P. Elbow, *Embracing Contraries,* Oxford, New York, 1986.
7. G. Baum, Forward to A. M. Greeley, *New Agenda,* Paulist Press, Garden City, New York, 1973.

8. A. Dulles, Model Five: Revelation as New Awareness, *Models of Revelation*, Doubleday, New York, pp. 98-114, 1985.
9. T. Fawcett, *The Symbolic Language of Religion*, Augsburg, Minneapolis, 1971.
10. B. Lonergan, *Insight: A Study of Human Understanding*, 3rd Edition, Philosophical Library, New York, 1978.
11. R. Young, A. Becker, and K. Pike, *Rhetoric: Discovery and Change*, Harcourt Brace, Jovanovich, New York, 1970.
12. A. Berthoff, *The Making of Meaning: Metaphors, Models, and Maxims for Writing Teachers*, Boynton/Cook, New Jersey, 1981.
13. E. Boyer, *College: The Underground Experience in America*, Harper and Row, New York, 1987.
14. D. Sloan, *Insight-Imagination*, Greenwood Press, Westport, Connecticut, 1983.
15. J. Piaget, *Six Psychological Studies*, Vintage Press, New York, 1968.
16. L. Festinger, *A Theory of Cognitive Dissonance*, Stanford University Press, Stanford, California, 1957.
17. J. Bruner, *On Knowing*, Belknap Press, Cambridge, Massachusetts, 1963.
18. L. Flower and J. Hayes, Problem-Solving Strategies and the Writing Process, *College English, 39*:4, pp. 449-461, 1977.
19. P. Tillich, *Systematic Theology, 1*, University of Chicago Press, Chicago, 1951.
20. W. Labov and D. Fanshell, *Therapeutic Discourse: Psychotherapy as Conversation*, Academic Press, New York, 1977.
21. J. Lauer, G. Montague, A. Lunsford, and J. Emig, *Four Worlds of Writing*, 2nd Edition, Harper and Row, New York, 1985.
22. M. Klein, P. Mathieu, E. Gendlin, and D. Kiesler, *The Experiencing Scale: A Research and Training Manual, 1*, Wisconsin Psychiatric Institute, Madison, 1969.
23. H. Strupp and J. Binder, *Psychotherapy in a New Key*, Basic Books, New York, 1984.
24. L. Shengold, Insight as Metaphor, *Psychoanalytic Study of the Child, 32*, pp. 289-306, 1981.
25. E. Fontinell, *Toward a Reconstruction of Religion*, Doubleday, Garden City, New York, 1970.
26. K. Rahner, Revelation, *Theological Dictionary*, Herder and Herder, New York, p. 411, 1965.
27. M. Belenky et al., *Women's Ways of Knowing*, Basic Books, New York, 1986.
28. B. Fisher, *Small Group Decision-Making*, McGraw Hill, New York, 1980.
29. A. Hare, *Handbook of Small Group Research*, 2nd Edition, Free Press, New York, 1976.
30. M. Shaw, *Group Dynamics: The Psychology of Small Group Behavior*, 3rd Edition, McGraw Hill, New York, 1981.
31. B. Burleson, B. Levine, and W. Samter, Decision-Making Procedure and Decision Quality, *Human Communication Research, 10*:4, pp. 557-574, 1984.

32. R. Cattell, Concepts and Methods in the Measurement of Group Syntality, *Psychological Review, 55*:1, pp. 48–63, 1948.
33. T. Scheidel, Divergent and Convergent Thinking in Group Decision-Making, in *Communication and Group Decision-Making,* R. Y. Hirokawa and M. S. Poole (eds.), Sage, Beverly Hills, pp. 113–130, 1986.
34. B. Collins and H. Guetzkow, *A Social Psychology of Group Processes for Decision-Making,* Wiley, New York, 1964.
35. M. Poole, Decision Development in Small Groups II: A Study of Multiple Sequences in Decision-Making, *Communication Monographs, 50*:3, pp. 206–232, 1983.
36. L. Putnam, Conflict in Group Decision-Making, in *Communication and Group Decision-Making,* R. Hirokawa and M. Poole (eds.), Sage, Beverly Hills, pp. 175–196, 1986.
37. M. Morgan et al., Collaborative Writing in the Classroom, *Bulletin of the Association for Business Communication, 50*:3, pp. 20–26, 1987.
38. R. Hirokawa and D. Scheerhorn, Communication in Faulty Decision-Making, in *Communication and Group Decision-Making,* R. Hirokawa and M. Poole (eds.), Sage, Beverly Hills, pp. 63–80, 1986.
39. B. A. Fisher, Decision Emergence: Phases in Group Decision-Making, *Speech Monographs, 37*:1, pp. 53–66, 1970.

PART II:
Case Studies of Collaboration

The value of case studies to technical communication research is well understood by scholars in the discipline. Case studies have greatly furthered our understanding of not only *how* writers write, but also *what* implications their writing and behavior may have within their respective organizations and disciplines. When researchers are able to gain access to organizational and other "real world" situations, they often provide significant case studies which offer other scholars in-depth pictures of writers and communicators at work. These case studies also create opportunities for the rest of the technical communication community to examine theoretical notions against observed behaviors. The efforts of such scholars as Selzer, Brown and Herndl, Doheny-Farina, Paradis, Dobrin, and Miller, Rymer, and Broadhead and Freed are representative of the valuable work which can result from researchers gaining access to organizations in which writers write [1-6]. This section of the collection continues this trend by providing case studies which offer researchers access to essential knowledge about industrial collaboration.

These case studies have certain similarities; for example, they both draw attention to interpersonal and management issues which pertain to collaborative writing. However, we should note that they have important differences as well; for example, one study was performed by observers-researchers who were not part of the organization they studied, while the other was done by a participant of the collaborative group under study. One focuses on dyadic interaction, while the other focuses on the development of group norms and practices as the group worked through their writing task. Nonetheless, both present an important picture of writers interacting with their colleagues, and both present suggestive avenues for further research and practice.

Barbara Couture and Jone Rymer in their "Discourse Interaction between Writing and Supervisor: A Primary Collaboration in Workplace Writing," explore writers' "interactive practices during composing." Prompted by results from their Writers' Survey, they suggest that collaborative writing in the "workplace might not be *group* writing as much as academic researchers have claimed but simply the writer interacting with others." They identify a particular dyadic relationship, between writer and supervisor, as accounting for the greatest portion of most writers' interaction with others. Their characterization of this interaction is important, both for current practice and for future research and pedagogy. Future researchers will particularly benefit from the review of their methodology used to conduct the research. From multiple case studies performed, here they select three which demonstrate how this primary interaction is largely "document driven." Moreover, Couture and Rymer draw attention to what the effects are when writer and supervisor hold differing perceptions of the purpose of the interaction. Finally, drawing implications from their case studies, they offer the reader some thoughts on which aspects of discourse interaction need further investigation.

Elizabeth Malone, in her "Facilitating Groups through Selective Participation: An Example of Collaboration from NASA," reports her observations about a problem-solving group at NASA brought together to write "some standard language" for important contractual documents. Leaders or managers of collaborative writing groups, she argues, can improve both process and product by recognizing and managing the various kinds of interpersonal skills group members possess and by employing those skills at appropriate times in the writing process. Her findings suggest that early phases of a collaborative writing project should be "expansionary," while later phases must, of course, move toward closure. Importantly, however, her case study demonstrates how different interpersonal skills and managerial techniques may be necessary at different phases of the collaborative process for it to be successful. She notes that collaborators need to realize that "the overall task includes phases where group members must contribute differently if the group is to succeed in its task." Her chapter concludes by discussing some implications for leaders or managers of group writing efforts.

REFERENCES

1. J. Selzer, The Composing Process of an Engineer, *College Composition and Communication, 34*:2, pp. 178–187, 1983.
2. R. Brown and C. Herndel, An Ethnographic Study of Corporate Writing: Job Status as Reflected in Written Text, in *Functional Approaches to Writing: Research Perspectives,* B. Couture (ed.), Ablex Publishing Corporation, Norwood, New Jersey, pp. 11–28, 1986.
3. S. Doheny-Farina, Writing in an Emerging Organization: An Ethnographic Study, *Written Communication, 3*:2, pp. 158–185, 1986.

4. J. Paradis, D. Dobrin, and R. Miller, Writing at Exxon ITD: Notes on the Writing Environment of an R&D Organization, in *Writing in Nonacademic Settings*, L. Odell and D. Goswami (eds.), The Guilford Press, New York, pp. 281–307, 1985.

5. J. Rymer, Scientific Composing Processes: How Eminent Scientists Write Journal Articles, *Writing in Academic Disciplines*, D. Jolliffe (ed.), Ablex Publishing Corporation, Norwood, New Jersey, pp. 211–250, 1988.

6. G. Broadhead and R. Freed, *The Variables of Composition: Process and Product in a Business Setting*, Southern Illinois Press, Carbondale, 1986.

CHAPTER 5

Discourse Interaction between Writer and Supervisor: A Primary Collaboration in Workplace Writing

BARBARA COUTURE
JONE RYMER

This chapter investigates a particular form of collaboration in workplace writing: discourse interaction between writer and supervisor. By discourse interaction we mean oral or written communication pertaining to a document during the process of planning, drafting, and revising it. Our research suggests that discourse interaction between the writer and other(s) may be the collaborative procedure most typical of professionals' writing in the workplace.

In the first stage of our research on collaboration, we surveyed over 400 professionals who write at work in a wide range of occupations and organizations. These employees responded to a five-part questionnaire, the Writers' Survey, which was distributed to professionals in the Detroit area, selected on the basis of their education, work experience, occupation, and competency as writers. The questionnaire elicited data on respondents' work experience, time spent composing particular types of documents, most typical writing tasks, and general style preferences. Among questions included on the survey were items inquiring about the frequency of writing or planning documents in groups and of interacting about writing for "routine" and "special" writing tasks. Writing was defined as routine when efficiency is more important than quality and special when quality is valued more highly. For both routine and special tasks, the professionals we surveyed only infrequently write with groups, but they typically interact with others during composing. We concluded that "collaborative writing" in the workplace might not be *group* writing as much as academic researchers have claimed but simply the writer interacting with others [1] .

Results from our Writers' Survey show a sharp contrast between writing with a group and interacting with others during the writing process. (The percentage of professionals who write or plan frequently with a group is compared in Figure 1 with those who claim to interact frequently with others about their special writing. Results represent the percentage of professionals answering "some," "often," or "very often.") For special writing, professionals claimed they infrequently write as part of a group or participate in a planning group for which they do the writing: only 24 percent contribute to a team-authored document sometimes, often or very often; and only 37 percent sometimes or more frequently do the writing for a group. On the other hand, the respondents claim to interact with others often during the process of writing at work: 76 percent said they sometimes, often, or very often "talk over their writing before drafting it;" 78 percent said that they sometimes or more often get feedback after drafting; and 81 percent said that they sometimes or more frequently revise on the basis of others' responses to their drafts. For routine writing, professionals claimed to engage in all tasks shown far less often.

We found these results from our survey research to be provocative. Collaborative groups might not be as ubiquitous as some researchers had suggested; yet interaction during the composing process seems very common. Perhaps academic assumptions about *multi-authorship* (with shared decision-making and responsibility), combined with the lack of an accepted definition of "collaboration," had obscured this point [1; 2, p. 70; 3, pp. 73-74; 4, p. 567] . And so our primary objective for a new study, whose initial findings are reported here, became exploring writers' interactive practices during composing.

The Writers' Survey results also surprised us in other ways and thereby defined our research agenda. Certainly, the prevalence of interaction during composing is striking, particularly when writers and/or their employers value the product highly and are concerned with quality more than efficiency (what the Survey designated as "Special Writing"). But oddly enough, writers in our survey claimed to interact with others as frequently *before* drafting their important documents as they do afterward. We wondered if this interaction in a pre-writing stage figures as significantly in writers' procedures as the interaction that takes place during revising. And we wondered about a great many other "how" and "why" questions, as well, because our survey results gave no

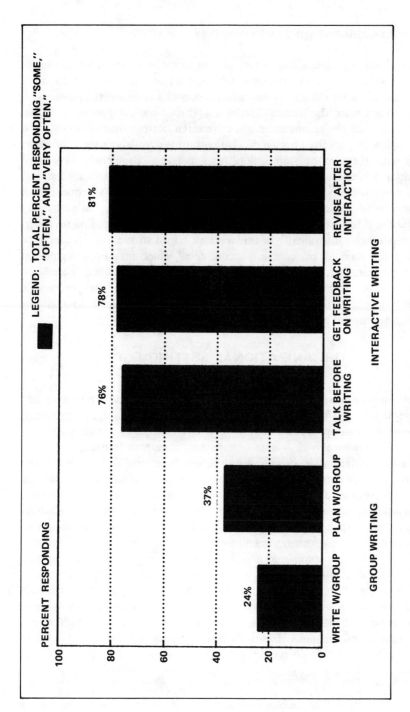

LEGEND: TOTAL PERCENT RESPONDING "SOME," "OFTEN," AND "VERY OFTEN."

Figure 1. Percentage of Professionals Who Participate in Collaborative Writing Groups Compared With Those Who Interact With Others While Writing

answers about the nature and quality of this interaction, about its motivating factors, about those who interacted with writers and the context for their interaction, or about the impact on writing processes and on written products.

The results from the Writers' Survey thus shaped our perspective, purposes, and methods for the second stage of our research. Rather than originating this study with a preconceived notion of what constitutes "collaboration," we began simply with the idea of "interaction" as a potential component in writers' composing processes. In this way, we hoped to develop a more detailed picture of the communicative context which characterizes composing at work. Rather than seeking to develop generalizations about common practices, our goal was to describe the practices of writers at work who interact with one or more others while producing a document.[1] In this way, we hoped to avoid the tendency to comment on obvious patterns and ignore detail which may prove significant upon close enumeration. Finally, rather than using empirical research methods, which tend to isolate and decontextualize, we conducted interpretive case studies—in-depth, holistic, and contextual explorations of interaction during composing. Three of our case studies are reported here.

ORGANIZATIONAL SETTING FOR THE CASE STUDIES

The organizational unit which we selected to research discourse interaction between writers and supervisors allowed contact with fully engaged respondents and constituted an appropriate setting to explore multiple case studies. The unit is an architectural service within a large health care institution, employing thirty-seven professionals, organized into several functional departments.[2] This in-house firm is representative of organizations participating in our original Writers' Survey, employing members of various professional groups requiring college degrees, whose jobs demand significant amounts of writing. A relatively new division, deep in the process of negotiating its communication procedures and devising reporting forms, the unit we studied has personnel who are highly engaged in discussing communication, questioning methods, and trying out new options. Hence, it offered us a rich communicative context where interaction could be assumed to be significant—a distinct advantage to our data collecting efforts. The unit's organizational structure also made it highly hospitable to collaborative writing; professionals are frequently assigned to project teams, often formed of members drawn from more than one functional area. In addition, the division produces numerous documents (especially proposals) that

---

[1] Although other composition researchers, notably Odell, have studied some aspects of discourse interaction in the workplace, our focus differs from his [5]. (Supervisor-subordinate communication is, of course, a large topic in management literature, but the focus differs radically from most composition specialists.)

[2] Some health care institutions have in-house architectural groups which can offer design and construction services for renovation and expansion demanded by advancing technologies and services.

require the input of a variety of stakeholders: it acts as a contract design group to users within the overall organization, as well as a liaison between the health care institution and its outside contractors and regulatory agencies. The situation is clearly a propitious one for studying collaboration.

We conducted multiple case studies, each involving a middle manager and a writer working for this architectural service, in a large public bureaucracy. The subjects, all architects or engineers, are the managers of a department within this service or a writer/supervisor working under the managers. All the available managers in this department participated in the study, and in cooperation with us, they selected the writers with whom they currently had the most frequent interaction.

RESEARCH METHODS

The essential characteristic about our methods is that they arise out of an "interpretive" framework. Positivist methods such as we used in the Writers' Survey provide broad generalizations about many writers' practices of writing collaboratively on the job—one kind of knowledge about a complex human activity. In this study we are concerned with another kind of knowledge, a deeper understanding of individual writers' experiences and intersubjective meanings created in specific contexts. Influenced by the "interpretive" approach and "social constructionist" theory of knowledge, our methodology is characterized by several major ideas: truth is created through discourse in specific contexts; observations are theory-laden; and the observer is critically involved with the observed. In short, our approach is guided by beliefs in: 1) individuals' perspectives on their social interaction; 2) intersubjective meanings achieved through interaction and revealed through interpretation; and 3) the human relationships developed between the researchers and the persons who are the "subjects" of study.

Rather than adopting one particular approach among the many in the interpretive movement, we have synthesized several, adapting them to what we learned in this specific context.[3] Such a flexible, non-standardized approach is typical of interpretive investigations, as Putnam explains:

> Although the exact goals and methods vary, interpretivists focus on the historically unique situation; they study naturally occurring phenomena; they become immersed or involved in the lives of the people they study; and they approach their task in a . . . flexible, iterative manner . . . [12, p. 44].

3 Our basic approach is influenced by many schools in the interpretive movement (such as naturalistic and qualitative inquiry, ethnography, case study, and hermeneutics), and by scholars across a wide range of disciplines, such as Kuhn, Polanyi, Rorty, Geertz, Fish, and others. (For discussions of the movement by composition researchers, see [6, 7].) On the application of interpretive methodology to discourse interaction in organizations, we are indebted to the work of Putnam, Morgan, and Lincoln and Guba [8-10]. For a demonstration and defense of the hermeneutic approach applied to functional texts, see [11].

Because we acknowledge the validity of completely different perspectives among those involved in discourse interaction [13, p. 864], our goal was not to search for one meaning or to generalize about the experience of many. Rather, we explored each interaction holistically. We hoped that our engagement in this process (our "collaboration" with those writers who are interacting over their writing in the workplace), would help us better describe individual experiences of interaction and explain intersubjective meanings in this particular context. The process also promised to reveal possible commonalities about interaction in the composing process, insights which may contribute to a theory of collaboration in the workplace. Finally, we hoped the potential of the process to reveal complexity would allow us to define hypotheses to test in other investigations.

We devised our methods, therefore, to help each of us as an individual researcher develop a thorough understanding of the perspective of one of the participants in each event involving interaction over composing—learning about the interaction and the document solely through his eyes. (All subjects in this study were men.) After each of us had embraced cognitively and affectively the view of one participant, we then, working as collaborating researchers, shared our knowledge and together tried to interpret the discourse interaction. Through our dialectic, we attempted to retain the messy details and to allow contradictions and conflicts to emerge in an effort to understand the interaction in some deeper way [11, pp. 104-105].

Although we used multiple techniques for data collection, the key methods, both bearing the imprint of the interpretive approach, were composing-interaction logs and discourse-based interviews. In both, we explored the perspectives of the individuals involved in the interaction without according them special status based on their roles; that is, we discounted the privilege academics typically accord to *writers* and the power that organizations confer upon *managers*.[4] Instead, we adapted these research methods to develop separate-but-equal constructs of each individual perspective involved in a composing interaction.

The Discourse-Interaction Log is a day-by-day record which directs the respondent to keep track of any communication he has concerning writing, both things authored by himself and those by others. The log focuses on a description of the participants' activities, but requests document type and the scene for the interaction. (A sample entry: "Edited a draft of a proposal [John] prepared and gave back to him to review . . . in a face-to-face informal meeting.") Discourse interaction includes both written comments and actual revising, as well as all oral interactions, even those by telephone, but explicitly excludes mere review and approval.[5]

---

[4] Managerial bias is characteristic of traditional communications research; eliminating it is frequently a goal of interpretive scholars [13, p. 39].

[5] Communication logs usually track only a writer's documents [14, pp. 16-17].

The purposes of the log were to develop a general framework for understanding document interaction in this context and to identify potential writers and interaction events for the study. To achieve these goals, three middle managers (heading architectural, electrical and mechanical engineering services within the unit) kept the log for a typical week, after which they were each interviewed. In addition, writers from each department were interviewed. Together with the log, these preliminary interviews allowed us to explore the practices and the patterns of discourse interaction in this unit from different perspectives and to select pairs of managers and specific writers. For each pair studied, we identified a recent (even on-going) document authored by the writer over which there appeared to be significant yet typical interaction and for which drafts were available.

Distributing the logs only to middle managers represented our strategy to locate significant composing interactions most efficiently. Because of their positions, these managers are in close contact with everyone in the unit and likely to be interacting frequently. Moreover, we anticipated that they would readily report document interaction with their writers. Our previous research led us to believe that people *other* than the writer are more likely to report document interaction; an author may dismiss comments others make about his writing or even the revising they do, conceiving of himself as "the one who did all the writing."[6] Whether true or not, each manager, though he reported drafting many documents himself, only claimed frequent and significant interaction about documents drafted by others. This finding may simply mean that managers interact about subordinates' writing more frequently than their own, but it may suggest that managers are more aware of the interaction when they are not the principal author and instead play a supervisory role. Although the logs, preliminary interviews, and the main interviews identified various "others" with whom writers interacted, the most substantive and frequent interactions were between writers and their immediate supervisors, the middle-level managers.

Discourse-Interaction Interviews featured questions about various iterations of documents to elicit respondents' detailed recall of interaction (on the general idea of Odell's "discourse-based interviews") [16]. We conducted these interviews as engaged interviewers, one of us individually with the writer and the other with the manager with whom he interacted. By limiting ourselves solely to the perspective of one participant in the interaction, we aimed to become fully involved and conversant with that individual's view, building up a separate, fully personal identification with his experience of collaboration with someone else. Rather than neutrally observing "what he claimed happened," we hoped to understand the individual's sense of the interaction, acknowledging that neither the writer nor

---

[6] Debs describes this phenomenon as respondents not recognizing discussion about writing as collaboration because of the highly collaborative context of the organization [15]. Also see Lunsford and Ede [3, p. 73].

the manager has the truth or "the facts;" instead, each simply has a legitimate personal perspective which we hoped to know from the inside, unswayed by the power of the manager's position or any assumptions about the authorial role.

Coupled with all of the respondents' keen interest in communication problems, this stance was particularly powerful in eliciting the confidence of the writers and thereby their fullest cooperation. Likewise, managers felt free to openly discuss writers' strengths and weaknesses, knowing we would not inadvertently expose this information to writers and thus jeopardize working relationships. While we promised each respondent that his information would not be revealed to others within the organization, our lack of knowledge about the situation and document, and our consequently naive questions, guaranteed this confidence.

Although we asked many questions about the respondents' general interaction about their writing, the focal point of each interview was the interaction over current documents—often in numerous versions. Focusing both the writer and the manager on a single, *current* communication problem helped to elicit rich responses about interaction during composing.[7]

Each of us independently wrote an extended description and analysis of our interviews prior to discussing them. These analyses, the interview transcripts, the logs, and the subjects' drafts and revisions of documents formed the data base for our subsequent collaborative interpretation.

THREE CASE STUDIES OF
DISCOURSE INTERACTION

For this chapter we have selected to report three of the case studies from our investigation of discourse interaction between writers and supervisors. The two writers involved (one of whom appears in two case studies) are similar in that both engage in frequent discourse interaction with their managers and both view themselves—and are viewed by their managers—as competent professionals but rather mediocre writers. The first two case studies pair the writer with the manager in their own functional work section; the third links one of these writers with the other manager, someone he reports to on certain projects.

In our discussion we first, describe the four subjects and their backgrounds, and second, present the three case studies. We treat the first two cases together, beginning with background information and following with our interpretation that suggests a theory of interaction during composing of "assigned writing." Our report of the third case study, which focuses on interaction over a single

---

[7] Recent psychological research on memory reinforces a common sense view that details about mundane events will not be recalled well beyond the immediate time frame. Focusing on recent coherent streams of events is most likely to elicit accurate recall [17, p. 161].

document, demonstrates the application of our theory. We conclude with implications for further investigation of discourse interaction in the workplace.

CASE STUDY SUBJECTS AND THEIR BACKGROUNDS

Our case study subjects include two managers, one of an engineering group and the other of an architectural group, and two writers, one from each unit. The managers, whom we will call Howard and Lewis, both confident about their own writing, are highly critical of their subordinates' writing skills. Manager Howard, in his late thirties, considers the writing of the four engineers in his section to be sloppy and inefficient, something he deals with by actually revising their work himself. Howard charges that some of their reports (e.g., internal monthly reports) are too detailed, whereas others (e.g., documents for outside clients) omit significant information. Although he admits that some problems occur because the unit is trying to establish its reputation (attempting to balance internal and external concerns), Howard believes his writers make repeated "mistakes" in their writing, such as forgetting to tell the reader what to do with a document. Howard says that he corrects these errors on their drafts, believing that these corrections reveal generalizations about "good" writing which his engineers can apply in preparing future assignments. Considering himself a very competent writer, Howard traces his confidence to a technical writing course and the influence of an early supervisor, who believed writing was all important and assumed an active role in commenting upon and editing Howard's writing.

Manager Lewis believes he should purposefully guide the writing of his engineers, but he spends most of his efforts reacting to their drafts, often rewriting documents to "make them sound the way I think they should sound." While he does "extensive," "diligent" monitoring of subordinates' writing, he believes he treats his co-workers fairly. Most of his interaction about their writing occurs at his weekly staff meeting, at which time members of his group show him drafts which he edits right then and there. Lewis in his mid-forties, thinks of himself as a good writer and "getting better," though not as good as he "would like to be." One of his major complaints (a prevalent one among all the managers) is that he doesn't have enough time to write, primarily because he is always behind in his supervisory work.

The two writers, whom we will call Yates and Roberts, express little confidence in their writing skills and are viewed by their managers as mediocre writers. In their managers' opinions, both men have difficulty perceiving the political implications of some of their writing assignments, of sensing how their writing might affect the design unit and influence its operational relationships with other units in the larger organization. Writer Yates, in his fifties, an electrical engineer in manager Howard's group, has been with the organization several years after many years' experience in industry. Writer Roberts, in his late twenties, reports to manager Lewis but occasionally works on projects under Howard. He has been with the organization a total of four years, taking out a year to work elsewhere.

Writer Yates admits that his writing often "misses the boat" on political matters; he claims to need someone else, usually manager Howard, to tell him how it would strike the readers. According to Yates, he typically takes drafts to Howard and stays with him for the document's review, so that, if he wishes, he can object to his manager's suggestions. Despite this adversarial posture, writer Yates claims they usually reach an amicable agreement. In addition to regular reviews with his manager, Yates occasionally seeks informal responses to his writing from a fellow electrical engineer in his group; interestingly this peer is viewed by manager Howard as the least competent writer in his section, at least a step below Yates. Writer Yates, however, finds his peer's reactions helpful, and Yates reciprocates by reviewing his peer's writing.

Writer Roberts is more confident about his writing abilities than Yates, but he too accepts supervisory corrections as a matter of course. Similar to Yates's interaction with Howard, writer Roberts claims his typical interaction with manager Lewis is over a drafted document where they democratically negotiate changes. Lewis typically changes much of what Roberts writes because Roberts has only recently returned to the organization and Lewis wants to avoid any possibilities of misrepresentation. In writer Roberts' opinion, Lewis revised his work when he first returned because it really needed it; now Roberts believes his writing has improved, but—out of mere habit—Lewis continues to revise. Roberts claims to take little offense at this pre-emptive activity, saying that English skills are not a particular strength among architects. Moreover, he believes Lewis has a way with words, citing the fact that the manager does all of his own writing, never asking others to draft for his signature.

Background on Cases One and Two

Case One deals with the interaction between writer Yates and manager Howard over Yates's assigned writing. Yates characterizes his discourse interaction with Howard as involving a lot of easy "back and forthing," some before beginning to write, but most after drafting. His perception of this interchange is that it is mutually beneficial, with both parties giving and taking criticism. His personal guideline for drafting is "to keep his supervisor out of trouble," but he admits he is not always able to do that without Howard's help. Nevertheless, writer Yates displays an easy confidence in his professional skills, and regards his reliance on supervisory guidance for his writing as only a minor problem.

Manager Howard interprets his interactions with Yates in quite a different way. His recollection of discussions over three recent documents all focused on his correcting Yates's "misunderstandings." In one instance, he claimed Yates did not understand that documentation was required to explain a cost overrun; in another, Yates reported far more technical detail to upper management than Howard thought necessary; and in a third, he failed to make it clear what action the reader was to take. Howard reports he's pretty hard on writer Yates for some of these transgressions and that the two yell a lot at each other. However, he insists that Yates never gets angry about his criticisms, and certainly he would never let him leave the office with either of them feeling mad. In short, this interactive situation, characterized as mutually productive and amiable by the writer, is perceived as somewhat punitive and one-sided by the manager.

Case Two explores interactions between writer Roberts and manager Lewis over assigned documents. Roberts claims that for typical assignments, Lewis gives him very little advice beforehand. Instead, Roberts produces a draft which they discuss in detail. He characterizes manager Lewis as a good listener who works with him to define the assigned project in the course of face-to-face interaction over drafts. He rarely sends out a document without Lewis's review, though he has done so when the document made "no real claims." On occasion, Lewis has expressed regret that a document was sent without his review because it "hurt the image" of the department.

Manager Lewis concurs that he spends considerable time editing Roberts' documents in his presence. He criticizes Roberts for not clarifying matters, often raising unnecessary questions in the readers' minds, and for making errors in grammar and punctuation. The manager notes that Roberts often fails to cite the causes of something, but he doesn't fault the writer for these omissions, attributing them to his inexperience and expressing confidence in Roberts' ability to learn.

Interpretation of Cases One and Two:
Toward a Theory of Discourse Interaction

Our interpretation of these cases one and two should contribute to a theory about discourse interaction in the assigned writing situation between managers and subordinates who are perceived both by themselves and their superiors to be mediocre writers. Two principal characteristics of discourse interaction in this "assigned writing" context emerge from our analyses:

1. Discourse interaction may be document driven, focusing on a pending draft rather than on pre-draft planning.
2. Writers and managers may have radically different perspectives on the function of discourse interaction, inhibiting them from collaborating effectively.

Our interview data confirms our previous assumption that managers and the professionals who work for them perceive interaction in the writing process to be functional primarily *after* a draft is written. This perception figured prominently in these three case studies, appearing to contradict our survey data which suggests that interaction about writing at work occurs as frequently before as after drafting. The answer may be that writers at work do frequently interact with others both before and after drafting, but that they perceive interaction after drafting to be more fruitful. The draft in hand presents the best opportunity for clarifying the terms of an assignment and/or for redefining a project, and it also creates a scene for negotiation and instruction about communication strategies. Our data also strongly suggest that managers and those whom they supervise may have different perceptions of the way interaction should best function in the composing process. The disparity in their perceptions, we believe, causes much of this interaction to be minimally effective.

The two participating managers believe interaction to be less fruitful before drafting because it takes too much time for too little return. Manager Lewis perceives it to be a worthwhile goal to interact with a subordinate about an assigned document before it is drafted, but too many other job responsibilities prevent that from happening. Ideally, Lewis believes that management should be more interactive and less "reactive," and that he should "not leave it up to employees to take a course or look in a booklet to get guidance." At the same time, manager Lewis admits that sometimes he can't judge beforehand what to tell an employee about a particular document and instead claims that his input should be supplying information—technical information which he can best add after he has seen a draft.

Lewis's hesitancy to assist a writer with preliminary assessments of what is required in a report appears to contradict his belief in an "interactive" management style. But his comments show he holds ambivalent beliefs about what he as a manager should do to guide his subordinates' writing. Even if conditions allowed him to interact beforehand with the writer who is completing a draft, he believes that such interaction would be a waste of his time. Most employees would not know how to make use of such a planning discussion; in the end, he would still have to revise the subordinate's draft to make it meet the conditions discussed beforehand, so no time would be saved. He believes only excellent writers could benefit from a planning discussion to clarify the assignment or discuss what the audience wants.

Since manager Lewis claims no professionals in his group qualify as "good writers," it is unknown whether he would, in fact, encourage interaction in the planning stage with such personnel. However, given Lewis's dissatisfaction with writers who continually misjudge audience expectations, one might expect him to attempt some preventative measures. Considering all of Lewis's comments, we suspect that either he does not know or cannot articulate what the situation demands and what he wants in a document until he sees a draft.

Both managers admit to some resistance to interaction with their writers for one other deeply held belief: "professionals" should not have to be told what to do. Claiming to echo the view of upper management, Lewis asserts that he and the employees under him are all professionals, and professionals should know their jobs, including how to write. Howard resists interaction before writing because he believes he has told employees time and again what to do by editing "mistakes" in their writing. Interaction beforehand, according to managers Howard and Lewis, should consist of the manager assigning a task to a professional who, because of his training as an architect or an engineer, should have adequate understanding to proceed on his own. There is little sense of this interactive activity being an opportunity for collaborative planning [18, pp. 68-75].

Writers Yates and Roberts also saw little utility in interacting about a writing assignment before a draft is written. However, they believe the draft to be a planning document—an opportunity for a discussion with their managers, clarifying the original assignment. Hence, what the managers perceive as mistakes to be corrected in their writers' drafts, the writers view as points to be negotiated as an assignment becomes more fully elaborated. Both of the writers believe that their draft in response to an assignment is a "trial balloon" sent up to see whether the manager will shoot it down. According to Roberts, interaction with his manager Lewis over a draft is the point at which they together actually work out the technical details of a project, such as overall cost, scheduling, and work commitments. Yates also believes his interactions with manager Howard over a draft are collaborative and democratic. Sometimes, however, he warns Howard beforehand "not to come down too hard on me."

Both writers believe the "trial draft" to be an effective way of clarifying assignments and finding out the most appropriate way to address them. They believe that their managers expect them to make a stab at their assignments rather than posing all their possible questions beforehand. They believe this partly because there are too many variables influencing a writing assignment which the manager knows and the writer does not; the writer cannot possibly address all contingencies beforehand.

Furthermore, writers use a first draft as a "trial balloon" because they may not have confidence in their ability to judge how well their writing meets an assignment. Yates, as we have described, often writes a draft and asks Howard to edit it because he is not confident that he can perceive how someone else may "take" his writing. Roberts judges that manager Lewis's changes in his writing often reflect his position as manager, especially on financial issues, and Lewis asserts he is careful to review anything where writer Roberts makes a commitment of money, time, or work. According to Roberts, Lewis makes changes for political reasons, adding content or changing quotes to "make the department look good" or "help sell something."

The managers of these writers also believe that interaction over a draft is an opportunity to communicate what is wanted, but they treat this communication as a "lesson" in which they are teaching their employees what to do and not do in the future. Both managers see interaction over a drafted assignment as an editing session. The manager edits the draft in the writer's presence and the edited changes "teach" the writer something about what to do next time.

In fact, manager Lewis claims his editorial changes "speak for themselves." He believes that it must be "obvious" to the writer why he makes the changes he does because they are "self evident" and the writers "don't argue" with them. Manager Howard claims to explain his changes orally, but he says he often just hands the document back to the writer, "letting them see" what they should have done. Aside from addressing the particulars of a specific writing situation in these interactions, Lewis and Howard claim to give some oral coaching in principles of good writing. Manager Lewis says he always tells his writers to be respectful of their readers' time and to write what people expect. Manager Howard repeats several "rules," such as pay attention to "secondary audiences" (a term he learned in technical writing courses), "think about what the reader wants to get from the memo," and "tell the reader what they are supposed to do with the document."

The writers believe that their managers use interaction over drafts to "teach" them, but they see the main function to be clarification of issues that could not be specified beforehand. For instance, in interacting with the manager, the writer may learn technical information he did not know or political considerations which affect the approach to audience. Both writers do claim to have learned some general principles to apply to their writing in these sessions. Writer Roberts believes he has learned from manager Lewis to depersonalize his writing, to omit "I" because it signals an opinion instead of a professional judgment. Writer Yates believes that he has learned from Howard to be "more political," avoiding language that may offend the reader.

Perhaps the most striking difference in the managers' and writers' perceptions of what goes on in their discourse interaction is their overall expectation. Although all consider the interaction sessions to be democratic, with a fair amount of give and take, the managers expect them to be primarily corrective, whereas the writers expect them to be informative.

Managers approach an interaction session over a draft intending to "correct" the writer's mistakes. Manager Howard, for instance, judges writer Yates's drafting "errors" as clear evidence that he did not understand the assignment. For example, Howard perceived that Yates misunderstood the function of a report form called a "Planning Investigation Report." Howard did not discuss the reporting requirements with Yates beforehand because he thought the report form made them obvious. When Yates submitted the report with a completed design rather than a feasibility study, Howard pointed out Yates's mistake, believing that the engineer had merely slipped into what he was accustomed to doing—cranking out designs. Writer Yates was disappointed at his boss's attitude, feeling that he should have been commended for taking initiative and presenting the design in the report, rather than merely doing a feasibility study.

Both managers wish that their writers could better anticipate what is needed in a document so that the manager wouldn't have to spend so much time correcting the subordinate's work. Although interaction before drafting might seem to be an obvious answer to the problem, neither the writers nor the managers share this perception. By and large, the writers accept supervisory revision as a fact of organizational life, sensing little of their managers' displeasure with the extensive revision tasks they regularly confront. For their part, the managers see little connection between lack of interaction beforehand and the extent of revisions required after drafting. In our view, manager Howard has developed a consistently critical perspective on his writers' work, blaming them for "faults" which he has identified and assuming no responsibility for their failures or their improvement. Manager Lewis, also quite critical of his subordinates, believes he should help them improve; the problem is that either he does not know or he cannot articulate what the situation demands and what he wants in a document until he sees a draft of it.

The patterns of interaction in each case study surely reflect individual participants' approaches to the collaborative writing process. But they also have features in common which corroborate a theory we wish to propose about the nature of discourse interaction in the supervisory setting.

The patterns of interaction we observed, though eventually leading to completed documents, were often described as ineffective by the participants. The interaction episodes did not lead to clear communication of writing assignments to subordinates, nor to clear definition of writers' problems in execution to managers. Both kinds of information, as we interpret the responses of our subjects, were expected of effective discourse interaction. The reporting relationship between supervisor and subordinate, exacerbated by the assigned writing situation, assumes pre-established roles of power and submission which may be antithetical to the cooperative effort required for effective interaction during composing, especially for collaborative invention. On the one hand, managers may equate their supervisory role with authorial responsibility and expect that writers will act as an extension of themselves, acting *for them* in the full sense of authorship; thus, for example, they may expect writers in their group to know as much as they do about an organizational problem and fault them for errors or omissions on a draft that are almost inevitable, considering the lack of supervisory direction about the assignment and the writers' limited perspectives of the situation. Yet on the other hand, writers who respond to assignments from their superiors may abrogate their responsibility for developing effective rhetorical strategies, simply assuming that their managers will do the job for them. In the process, such professionals fail to match the full technical responsibility they are expected to assume as architects or engineers with full authorial responsibility in reporting.

Background and Interpretation of Case Three

To illustrate the disjunction of managers' and writers' perceptions about interaction over an assigned document, we will report next a retrospective account of the interaction between Roberts and Howard over a draft (See Figure 2). The typed draft was composed by Roberts; manager Howard's corrections are handwritten. The document presented two rhetorical problems for the writer (whose technical task was to specify safety locks for a laser treatment room): first, the letter had to convince the addressee, a fire marshall, that the locks meet fire safety specifications; and second, the document required managerial authorization. Manager Howard perceived that he had advised Roberts that the letter was to be signed by another manager, Kip Ewing; Roberts initially presumed he himself to be the signatory and may have believed the issue was negotiable. Manager Howard's "corrections" to the document in the main involves changing the signatory yet again (finally signing it himself and copying Ewing), as well as adding technical details.

Manager Howard's and writer Roberts' reports of interaction prior to drafting focus on the assignment. Howard believed he clarified the signatory and assumed Roberts would know how to draft the letter from the perspective of Ewing, another manager, and was most disappointed by his presentation of self (e.g., writer Roberts' use of "I"). Howard did not believe the issue of the signatory to be negotiable, nor did he believe he had first named Roberts as the signatory. Furthermore, manager Howard did not see his post-draft decision to change the signatory from Ewing to himself as confusing; in his view, Roberts should have known how to write from the perspective of a manager, whether the signatory were himself or someone else.

Writer Roberts had a very different perception of the assignment; he believed that it was customary for such a document to be signed by himself and that only after the fact did Howard indicate that Ewing was to sign it. Roberts reports that at the time of drafting, he believed himself to be the designated signatory. However, the fact that in this draft he refers to himself as "I" and yet names Ewing as the signatory may reveal some confusion about the designated authorship.

In discussing their interaction over the draft, Howard confirms the managerial perspective that the interaction involves correcting the writer's mistakes while Roberts confirms the writer's perspective that the interaction involves an opportunity to clarify the assignment and ascertain managerial objectives. Howard perceived Roberts' choice to state his case in the first person as if he were the signatory as a mistake. He claims that Roberts gave him the draft, commenting that he did not know how to write it as if it were coming from Ewing; manager Howard viewed this as a lazy attempt to pawn off the task of revising the document. Writer Roberts, however, viewed his choice to state things in the first person as a deliberate rhetorical move: to him, the situation

[Hospital Letterhead]

Mr.
Department of State Police
Fire Marshal Division
General Office Building

Subject: Control of Access Doors into Laser Treatment Room
Project: Yag Laser Installation
Project No.: 2M 88 001

Dear Mr.

a follow up to our telephone conversation on Thursday, September 3; 4:30 p.m. I am sending a sketch of the Treatment/Control Room. We discussed the possibility of installing a thumb controlled latch at the control booth door leading to the staff corridor. In the interim it has come to my attention that We are also proposing an electromagnetic lock at the Treatment Room door to the patient corridor. Both doors and the proposed locking mechanism are identified in the attached sketch.

We would appreciate your review and comments regarding both doors. For clarity I have identified the two separate sets of issues as Issue A and Issue B.

Issue A: Installation of a thumb controlled latch at the Control Room to staff corridor door.

Purpose: This latch will prevent accidental entry from the staff corridor, during treatment procedures. This will protect Hospital staff from accidental exposure to the laser. The Control Room can safely be entered through the adjacent Scrub Room. The Scrub Room would also allow for access from the Control Room in the event of an emergency.

Issue B: Installation of an electromagnetic lock at the Treatment Room door to patient corridor.

Purpose: This lock will prevent accidental entry from the patient corridor during laser treatment. I have provided a manufacturers description of this lock for your information. This lock would be wired into the equipment wiring for the Yag Laser. When the laser is activated this lock will prevent entry until the treatment is complete. As a safety feature this lock will also be wired into the hospital alarm system so that the lock will be deactivated and the door would be free to open.

Figure 2. Safety Lock Document: Sample of Discourse Interaction

Your review and comments will be appreciated. Should you require any further
information on this project, please do not hesitate to contact ~~George Roberts~~ Design
Manager at [telephone]. Jim Howard

~~Sincerely,~~

approved by:

Kip C. Ewing
Director, Capital Acquisition Management Office

GR/BLD
1-gr98

cc Def

Sincerely

James Howard
Manager, Electrical Engineering

Figure 2. (Cont'd.)

seemed mundane, not one requiring a manager's signature. Roberts also claims that Ewing was added as signatory in his interaction with Howard after he drafted the document, a point that Howard denies.

Finally, the manager and the writer differ in what they feel should be the manager's responsibility in revising the writer's draft. Editing which involves corrections related to each's perception of the writing problem is viewed as unacceptable by the manager and unavoidable by the writer. Howard felt that Roberts should have understood the rhetorical situation beforehand; the writer felt the manager was undecided about the authorship of the memo, and that the issue was negotiable. However, editing which involves addition of technical content unknown to the writer is acceptable. Roberts, an architect, felt dependent upon Howard's expertise as an engineer for the technical specification of the laser lock; Howard, correspondingly, thought editorial changes related to the lock specifications were appropriately his responsibility.

TOWARD FUTURE INVESTIGATION OF INTERACTION OVER WORKPLACE WRITING

Our observations of the interaction about writing between managers and the professionals who work for them raise questions about the nature of that collaboration for future research. First, we need to know more about the extent of interaction about a writing assignment before it is drafted. Our interview data suggest that extensive planning involving the technical problem addressed in a document may take place before drafting, but that managers and their employees may neglect to take advantage of the opportunity for such discussion about rhetorical problems and strategies. Not to be overlooked, of course, is the link between management style and elaboration of the writing assignment. Our data suggests that managers view collaborative planning over technical issues to be appropriate, but collaborative planning over rhetorical issues to be unnecessary.

The lost opportunity for collaborative planning in the assigned writing situation is particularly significant because we know from our previous research that professionals are predisposed to planning their communications. Our survey research confirmed that for special writing, 90 percent of a population of over 350 writers mentally plan before drafting special writing. Furthermore, 98 percent of these professionals jot down notes or lists before drafting special writing at least some of the time, and 60 percent frequently outline their special assignments [19]. We believe that a constructive approach to collaborative planning might make professionals' composing processes more productive.

Second, we should further investigate the function of a draft in negotiating the terms of a writing assignment. In our interviews, the writer's draft was clearly the focus of all interaction—an impetus for collaborative planning. In the supervisory situation, a clash between the manager's and worker's views of the function of this negotiation appears to make it less satisfactory for both. Our assessment of the critical importance of collaboration after drafting in the assigned writing situation is confirmed by other research. For example, in a report of writing practices at Exxon ITD, Paradis, Dobrin, and Miller note the prevalence of "document cycling," that is, the process of a piece of writing moving back and forth between writer and supervisor, with attendant corrections from the supervisor and revisions from the writer. The researchers interpret the process as "a collaborative, if sometimes, stormy, process of managing work" [20, p. 294]. The cycled drafts become the scene for asserting managerial authority as well as for redefining projects, much as in the cases we reported here. In the Exxon study, writers reported being confused about the purpose of the cycling process; our own study demonstrates that managers and writers may have vastly different perceptions about the function of discourse interaction. Future studies should further investigate the process of document discussion and review and its potential for improving communication in the workplace.

Finally, we believe our case study shows that revision with others may be a critical feature of the assigned writing situation, a feature that may be overlooked in studies of professional composing that omit records of interaction with others (e.g., Selzer's report of the composing process of an engineer which depicts an individual's process of highly structured planning and little subsequent revision [21]). For the professionals we observed, revision was instigated in the main as a result of interaction with managers who assigned the writing.

In addition to these conclusions, we have learned from this initial analysis that retrospective interviews of interaction over writing are a powerful data base for interpretive analysis. They can reveal much about writers' perceptions of the utility of that interaction in the composing process that may not be gathered through protocol analysis or other methods of studying composing. Interpretation of such data should prove valuable as we work to help writers use interaction more effectively in the process of composing.

REFERENCES

1. B. Couture and J. Rymer, Interactive Writing on the Job: Definitions and Implications of "Collaboration," in *Writing in the Business Professions: Research, Theory, Practice,* M. Kogen (ed.), National Council of Teachers of English/Association for Business Communication, Urbana, Illinois, pp. 73–93, 1989.
2. N. Allen, D. Atkinson, M. Morgan, T. Moore, and C. Snow, What Experienced Collaborators Say About Collaborative Writing, *Journal of Business and Technical Communication, 1*:2, pp. 70–90, 1987.
3. A. Lunsford and L. Ede, Why Write . . . Together: A Research Update, *Rhetoric Review, 5*:1, pp. 71–81, 1985.

4. L. Faigley and T. Miller, What We Learn from Writing on the Job, *College English*, *44*:6, pp. 557–569, 1982.
5. L. Odell, Beyond the Text: Relations between Writing and Social Context, in *Writing in Nonacademic Settings*, L. Odell and D. Goswami (eds.), The Guilford Press, New York, pp. 249–280, 1985.
6. P. Bizzell, Foundationalism and Anti-Foundationalism in Composition Studies, *Pre/Text*, *7*:1-2, pp. 37–56, 1986.
7. K. Bruffee, Social Construction, Language, and the Authority of Knowledge: A Bibliographical Essay, *College English*, *48*:8, pp. 773–790, 1986.
8. L. Putnam and M. Pacanowsky (eds.), *Communication and Organizations: An Interpretive Approach*, Sage Publications, Beverly Hills, California, 1983.
9. G. Morgan, *Beyond Method: Strategies for Social Research*, Sage Publications, Beverly Hills, California, 1983.
10. Y. Lincoln and E. Guba, *Naturalistic Inquiry*, Sage Publications, Beverly Hills, California, 1985.
11. S. North, Writing in a Philosophy Class: Three Case Studies, *Research in the Teaching of English*, *20*:3, pp. 225–262, 1986.
12. L. Putnam, The Interpretive Perspective: An Alternative to Functionalism, in *Communication and Organizations: An Interpretive Approach*, L. Putnam and M. Pacanowsky (eds.), Sage Publications, Beverly Hills, California, pp. 31–54, 1983.
13. W. Swann, B. Pelham, and D. Roberts, Causal Chunking: Memory and Inference in Ongoing Interaction, *Journal of Personality and Social Psychology*, *53*:5, pp. 858–865, 1987.
14. D. Goswami, J. Redish, D. Felker, and A. Siegel, *Writing in the Professions: A Course Guide and Instructional Materials for an Advanced Composition Course*, American Institutes for Research, Washington, D.C., 1981.
15. M. Debs, The Technical Writer and Corporate Influence, Paper read at Thirty-seventh Annual Conference on College Composition and Communication, New Orleans, March 13–15, 1986.
16. L. Odell, D. Goswami, and A. Herrington, The Discourse-Based Interview: A Procedure for Exploring the Tacit Knowledge of Writers in Nonacademic Settings, in *Research on Writing: Principles and Methods*, P. Mosenthal, L. Tamor, and S. Walmsley (eds.), Longman, New York, pp. 221–236, 1983.
17. N. Bradburn, L. Rips, and S. Shevell, Answering Autobiographical Questions: The Impact of Memory and Influence on Surveys, *Science*, *236*, pp. 157–161, April, 1987.
18. K. LeFevre, *Invention as a Social Act*, Southern Illinois University Press, Carbondale, Illinois, pp. 68–75, 1987.
19. B. Couture and J. Rymer, Professional Writing and Situational Exigence: Profiling Writing on the Job by Task and Profession, in *Writing in the Workplace: New Research Perspectives*, R. Spilka (ed.), Southern Illinois University Press, forthcoming.

20. J. Paradis, D. Dobrin, and R. Miller, Writing at Exxon ITD: Notes on the Writing Environment of an R & D Organization, in *Writing in Nonacademic Settings*, L. Odell and D. Goswami (eds.), The Guilford Press, New York, pp. 281–307, 1985.
21. J. Selzer, The Composing Processes of an Engineer, *College Composition and Communication, 34*:2, pp. 178–187, 1983.

CHAPTER 6

Facilitating Groups through Selective Participation: An Example of Collaboration from NASA

ELIZABETH L. MALONE

Any problem-solving group experiences a tension between the demands of the group norms and the demands of the task. Where the group tends to set up for itself ways of operating, goals, and values that remain the same, the task demands different ways of operating, goals, and values at different stages of the problem-solving work. A group whose norms work well at one stage of a task may find itself stymied by those same norms at another stage of the task.

On the one hand, a problem-solving group (especially when the members know each other) tends to quickly establish a normative consensus when it is formed and tends to cling to this consensus throughout the task at hand. Hopkins defines the normative consensus as "a set of ideas and sentiments to which the members are motivationally and morally committed" [1, p. 45]. Under the established consensus, members behave in the same ways every time the group meets. On the other hand, the overall task includes phases where group members must contribute differently if the group is to succeed in its task. The early phases of a task should be expansionary—seeking out all possible ways of looking at and solving the problem. Spontaneity and receptivity should characterize a group looking for creative and optimal solutions. Later phases of a task need to be more narrowly focused; that is, the group needs to choose a solution, to see and present the chosen solution in depth, and to attain the closure of having completed the document.

The tension between the group's normative consensus and the changing demands of the problem-solving process means that individual behaviors—and indeed the normative consensus—may be productive for the group in some phases of the process, but counterproductive in other phases. On the level of individual roles, for example, the group may include a person who habitually questions and tests every proposition made by other group members. One group's patterns of behavior may include Jeanine saying, "I don't see the purpose of that. . . ." Though Jeanine's behavior will tend to remain the same, her constant questioning will be blocking behavior when the group is trying to generate ideas, brainstorm, or otherwise expand its realm of thinking. However, that same questioning behavior may be extremely helpful as ideas are tested, discussed, and chosen. Conversely, if Bob is the person who continually comes up with new ideas, he will be invaluable in the brainstorming phase, but counter-productive in the decision phase. On the level of group norms, if a group norm states that all decisions must be unanimous, the group cannot function when real differences occur [2].

This chapter examines this tension more closely, then discusses it in the context of a specific collaborative writing group. The discussion revolves around the group as a system with its own ways of behaving, developing, and producing. The group as a system is more than a collection of individuals or behaviors or roles; the group itself exhibits certain behaviors and has certain goals which may not coincide with individual goals [3,4]. As Fuhriman, Drescher, and Burlingame have pointed out, most group research has focused narrowly on single topics about groups, tackled the issue of the individual versus the group, or assumed that all small groups act in uniform ways [5]. This discussion focuses on the overall process of the group and how that process can be enhanced.

TENSION BETWEEN GROUP NORMS AND
THE PROBLEM-SOLVING PROCESS

Groups often form and reform around specific tasks such as writing a document. Generally such groups are called task groups or problem-solving groups. They have in common their ephemeral nature and their dedication to one (usually fairly defined) task, and the task will usually have an ending point, that is, a deadline. Examples of these kinds of groups include proposal-writing groups and groups who will give major presentations. Group members may know each other very well, slightly, or not at all, but generally they will be in this configuration for this task only; for example, a proposal-writing group will consist of some or many of the same people from one proposal to another, but there will usually be some change of membership depending on the work and types of costs being proposed.

The first group meeting may set up most of the group norms, especially if the group is narrowly focused on a short-term task and especially if most group members already know each other fairly well. Unfamiliar members will be introduced to the group, and the group will gain some idea of the expertise each brings to the group. Often the designated group leader (perhaps the Proposal Coordinator, in the case of the proposal-writing group) will explicitly allot to each person in the group a share in the whole, but the assignments will be task oriented, for example, what sections of the proposal will be written by the engineer, the business manager, and the proposed project leader. Most group leaders will not make statements like, "Jill, we'll look to you to keep us on track; Bob, you'll lead in coming up with new ideas; Will, you make sure that everybody is polled for opinions, so we have consensus." However, after the first meeting, group members usually have unspoken assumptions about who will perform those roles and have probably already performed in the roles they will continue in; for example, Will has already asked quiet Marcia's opinion on some issue, and so on. The group has gone through the complex process of sizing each other up, determining how each member fits (or does not fit) into the whole group, and taking on individual roles to play for the duration of the group. Group norms will also be set up, usually informally as members set up the process and the goals of the group. At the second meeting, members will perform their individual roles and group norms will be well established, the group norms setting standards for what must be done when, the individual roles establishing who does the things that fall within the norms [6]. Norms and roles may extend to the physical setting, as well as to behaviors. If the meeting is held in the same place, members will likely sit in the same places as they did for the first meeting and eye contact paths will remain the same (whether or not that was effective for each member).

Both positive and negative roles or blocking behaviors will be quickly established within the task group. If Art failed to participate in the first meeting, he will likely not participate much in any future meeting. If Bob did not contribute new ideas in the first meeting, he may not contribute any (or many) during subsequent meetings, even if that is his usual role in groups. If no agenda-setter came forward at the first meeting, future meetings may go on and on without achieving resolution. If most group members felt uncomfortable with the task assignments or the *modus operandi* at the group's initial meeting, the comfort level will probably remain low. If the group leader dictated an unpopular norm, such as allowing no joking, group members will continue to feel antagonistic toward the leader and, possibly, toward the group.

Indeed, a group may be defined in relation to its group norms and roles. According to Hopkins, the group is a group, based on 1) the degree of consensus on normative expectation; 2) the degree to which interaction occurs; 3) the degree to which interaction occurs in accordance with binding expectations; 4) the degree to which members and others define the group as a group; 5) the degree of visibility of group norms; and 6) the extent to which members evaluate each other by the expectations they hold [1, pp. 11-13]. Conflict about group norms will make the group "less of a group;" too much unanimity is possible, too, when conforming to the norms becomes more important than any of the group's members or the group's activity [7].

The group *qua* group quickly establishes and maintains its norms, with members performing established roles and the group taking on a personality of its own. However, the demands of individual phases of the task call for different skills, indeed, different *types* of skills, from the group as a whole. Following is a list of standard phases most problem-solving or task groups go through to complete their work:

1. *Defining the problem.* Sometimes this is done in advance by the "group leader" or task-giver; sometimes it is not done at all. In most cases, the general problem is defined in the first meeting but redefined by the group as it sets about solving the task, for example, as the group discovers that the document has additional readers.

2. *Creating criteria.* (Here is a key point at which the task or purpose is often redefined.) Here the group tries to think of as many ways to think about the problem in as many different ways as possible.

3. *Generating possible solutions.* Usually the group uses some form of brainstorming, nominal group process, or other idea-generating mechanism to gain as wide a range of possibilities as is feasible.

4. *Measuring possible solutions against criteria.* Ideas are scrutinized, redefined, combined if necessary, and evaluated to see if they are "still in the running" to be part of the final document or solution. One idea may emerge as a clear "winner," or the group may decide to present several alternatives in the final document.

5. *Codifying the solution, often by a final report or by establishing a first/ final draft of the document the group was assigned to write.* One group member may write this and present it for other members' review and comment, or the group may generate the document by giving parts to individual members or writing in group sessions.

The process outlined above is not a straight-line process; rather, it tends to be recursive, both in its broad outlines and for individual pieces of the task. For example, if the problem or task gets redefined, then the group's best interests would be served by repeating steps 2 and 3, at least in part, since a redefined problem would call for redefined possible solutions. Or the group may think it has consensus on the best solution, but find much disagreement as writing the final report gets underway; members may go back to steps 3 and 4 at that point.

Both this process and its recursiveness challenge a group which has established group norms to break out of those norms in order to function in different ways. The expansive, idea-generating skills needed for brainstorming and establishing criteria are very different from the analytic, closure skills needed for measuring ideas against criteria. When consensus is reached, the group may need to go through a parallel process in order to generate a report; as a *writing* group, members need to define the communication problem, create criteria, and so on. This process will again demand a shift in types of skills during the process. How can a group engineer the needed changes of skills and styles in order to accomplish its task?

A skilled leader can engineer such changes, usually by evoking the skills of different individuals at different times in the process. Bob, for example, will find he is facilitating the brainstorming session, while Jeanine sits more or less silently; she has her turn leading the sessions where possible solutions are measured against criteria. One college professor never said a word in his small seminars but controlled them using eye contact. When the seminar was getting off track, he would look steadily at the person who always asked what *that* had to do with the issue at hand; when the group was stagnating, the professor would look steadily at the person who generally came up with new ideas; and so on. A skilled leader can control and direct the process, capitalizing on all the individual strengths of the group's members.

Similarly, if the group members themselves are skilled at small group processes, they can themselves take on different roles at different stages of the process. However, group leaders and members who are highly skilled and adaptable are the ideal, *not* the reality in most workplace settings.

Of course, with or without a skilled leader or skilled members, changes in groups happen all the time; however, these changes are often not made for the specific purpose of enhancing the group's effectiveness. In fact, most often any changes are matters of pure exigency. Sometimes circumstances will change; even a change in the meeting place can affect group norms and roles. However, most often, and most importantly, group membership will change, as individual members change jobs or move to higher priority tasks. These changes in membership are most likely to have meaningful effects on how the group operates. Since these changes tend to occur anyway, perhaps groups or group leaders could direct such changes in order to make the groups more effective. This is what happened in one collaborative writing group at NASA Headquarters.

AN EXAMPLE OF COLLABORATIVE WRITING AT NASA

A NASA Employee Team (NET) was formed to write some standard language for the Requests for Proposals (RFP's) generated by the Contracts and Grants Division. The Division's government managers felt that Section L, "Instructions to Offerors," and Section M, "Evaluation Factors for Award," should include much the same language for each Request for Proposals. The format instructions could certainly be the same, and the categories of information would vary only slightly, depending upon the type of work NASA would be contracting for. Whether the RFP called for supplies, support services, or research, the government would still want offerors to demonstrate their understanding of the requirement, include information about their key personnel, and discuss proposed management of the contract.

The NET, then, was supposed to generate several different versions of Sections L and M, which could then be incorporated into individual RFP's and, if necessary, tailored to individual requirements. The Division's managers asked the Systems Branch Chief to form the NET, over his protests that he could generate the standard sections faster by himself. The Branch Chief chose:

- his Policy and Review Officer (PARO), who, as reviewer of all RFP's, was well aware of the problems of having the sections written "from scratch" or by piecemeal use of prior sections;
- a Senior Negotiator who was conducting a very large acquisition;
- the Senior Negotiator who had been at NASA Headquarters the longest time (eight years); and
- a Technical Writer.

Two aspects of group membership should be noted here: first, that the two negotiators represented different acquisition branches (the third branch declined to participate in the NET); second, that all NET members were government employees except the Technical Writer, who was a contractor employee.

Initially the Systems Branch Chief hoped to use the NET as a kind of reviewing group. He would generate the sections, the NET would review, revise, and edit, and the sections would be done, probably, in a month or so. The Branch Chief carried out the first part of this plan by giving all group members fairly complete versions of Sections L and M for support services RFP's before the first NET meeting. Then, in the first meeting, he presented the categories of L and M versions he thought would be appropriate: one version for support services, a second for research, a third for supplies, and a fourth for all non-competitive actions (that is, when an RFP is issued to only one contractor, much of the language intended to ensure fairness for all offerors is superfluous). He then asked for comments on his initial drafts.

However, it quickly became clear that the other government NET members did not see themselves in reviewing/revising roles. Though they accepted the Branch Chief's four categories, they rejected his draft and explicitly decided to hold meetings for each version of L and M, bringing examples to the meeting and writing text in the meeting. The Branch Chief objected strenuously that this process would be extremely time-consuming; the government NET members responded that they felt this would be the only way to get a true consensus document. (The Technical Writer stayed fairly quiet throughout this discussion, since she had reasons for supporting both positions. On the one hand, she was officially assigned to the Systems Branch Chief; on the other hand, she felt that the group process would, indeed, produce a better document.)

The Branch Chief accepted the group's decision, though he clearly felt uncomfortable with it. By the end of that initial meeting, group norms were well established. Some of the more important of these norms were:

1. The document would be a "grassroots" document, created by NASA line staff for NASA line staff, not created by management and forced on line staff.
2. The document would draw from what had worked well in the past, from actual examples of RFP's where Sections L and M facilitated the evaluation process. The document would draw from as many of these sources as possible.
3. It would be better to include too much and cut later than to include too little and force people to come up with other text on their own.
4. The deadline was important, but not as important as having a good document.

The Systems Branch Chief continued to contribute text in the month that followed the first NET meeting, and the group continued to use his text as only one example. Each government member brought other examples, and more text was incorporated into the NET's document from other examples than from that of the Branch Chief. The Branch Chief, impatient with the group process and worried about the deadline, understood that his contributions were being seen as blocking the process and informally withdrew from the group for about two months, though the NET continued to meet in his office. (He simply excused himself from each meeting as it came up by pleading a conflict.) Everyone in the group understood his motivation, and, in fact, he talked about it openly, making comments such as, "You guys are doing fine without me." However, he continued to ask the members how the task was going, and he asked the Cost/Price Analyst, one of his staff members, to contribute to the subsections which dealt with cost proposals.

The Cost/Price Analyst, a "new" member of the group, reinforced the group's expansionary and inclusive tendencies. Other members had the general goal of getting sufficient cost information from offerors; the Analyst produced a set of nine cost exhibits, one of which had up to six parts. (In the review stage, Acquisition Branch Chiefs and the Division Director objected to having so many cost exhibits; eight months later, several of the exhibits were eliminated.) The group went over each exhibit with the Cost/Price Analyst and included all of them in the standard "Instructions to Offerors." Before and after (but especially after) the two meetings that the Cost/Price Analyst attended, the NET treated other subsections in the same way, tending to include as much as possible. From three examples, the group would choose the most inclusive, then add text from the other examples. The Technical Writer would draft the subsection—and redraft it several times as more text was added.

During the period of the Branch Chief's absence, the NET became more and more expansive, including everyone's ideas and examples, generating draft after draft, and continually refining each draft. Everyone's opinion was welcomed and utilized; different approaches were set side by side and aspects were drawn from each to create still different approaches. Although all members had fairly heavy schedules, they managed to meet almost every week for two hours and to make time to review drafts between meetings.

However, after about six weeks the NET seemed to reach the limit of its expansionary activities but did not know how to begin attaining closure. The group rule that everything should be considered and, if possible, included began to block attainment of the final step of the process: establishing a text. Meanwhile, the group's May 31st deadline moved closer.

The Systems Branch Chief became concerned about the seeming lack of progress (that is, toward final text) in the group and asked the Technical Writer to intervene. She attempted to establish subsections as final by asking group members to sign a cover sheet for each subsection, explaining that their signatures would mean this draft would be the final draft. However, although members signed, they clearly did not see the versions as final. For example, even when one negotiator said he was going to read the "final" version only for typo's, he had half a dozen substantive changes to suggest. The Technical Writer explained the situation to the Systems Branch Chief. He decided (probably independent of this explanation) to begin attending NET meetings again.

His reappearance changed the membership of the group and, consequently, a change in group norms was possible. Furthermore, group members were willing to change, realizing that the group was getting bogged down in endless rewrites but was unable to change its norms. The Branch Chief reintroduced his first text, but this time as a vehicle for asking hard questions about the end purpose of the document. He also showed support of the principle of inclusiveness while reassuring the group that everything worthwhile had been included. He further supported the group's evolving decision to generate only two versions of Sections L and M—one for competitive actions of all kinds, one for non-competitive actions.

The subsections still needing major work when the Systems Branch Chief re-entered at the beginning of May were those dealing with the Technical Proposal. For these subsections (in both Sections L and M), the combination of the old group norm of inclusiveness, the new group norm of testing text against end purpose, and the approaching deadline helped the group to generate final text in three productive sessions. The final text was sent out for review on May 26th.

Even during May, however, the group had misgivings about finalizing text, about letting go of the text, and about the possibility that more items should be included. Several times group members had to explicitly tell each other that 1) the text was not "set in concrete" but could be changed as time and experience indicated needs for change and 2) negotiators could tailor the text to individual Requests for Proposals if it was not suitable "as is."

This balance the group had achieved under the leadership of the Systems Branch Chief was maintained during the review stage. Three of the five reviewers had substantive comments to make on the twenty-four-page document, and the group had to consider whether to include suggested changes, reject changes (which included attempts to persuade reviewers that rejection was the right thing to do), or reconcile conflicting comments in some way. The older group norm moved the group in the direction of including all comments and suggested changes, which was obviously impossible. This group norm, as reinterpreted by the addition of the Branch Chief, tried for inclusiveness but recognized that some inclusions would be counter to the purpose of the document. So when the Technical Writer consolidated the reviewers' comments, group members could see immediately that some changes were just not going to be made. Again, though, the group had to reassure itself that changes could be made in the future and that specific acquisitions would require changes to the standard language.

ANALYSIS OF THE GROUP PROCESS

The success of this collaborative group hinged on the perceptive, selective participation of the Systems Branch Chief. Initially, the Branch Chief recognized that his preferred *modus operandi* ran counter to that of the rest of the group. After making an honest attempt to get the group to adopt his process, he saw that his continued presence was blocking the expansionary activities of the group as a whole. He then judiciously withdrew, though keeping contact with most group members outside the group sessions. In this way he was able to keep a place for himself in the group and actually facilitate the group process by his absence. He understood himself well enough to realize that he could not adapt his own ways of thinking and operating to the group's; he had enough self knowledge to know that he was not the ideal group leader. However, he had enough confidence in the abilities of the group members that he could let the group function without him, at least for a time. He held himself ready to re-enter the group and did so when he saw a fruitful role for him to play.

The NET's recursive writing process became unproductive at the stage where the group needed to generate a final document. The group norm of inclusiveness, established at the first group meeting, left the group going around in circles of drafting and redrafting without reaching a final document. At this point, the Branch Chief's "closure" skills—initially a blocking mechanism in the group's process—became exactly what was needed to change the group norms so that a final document could be produced. His ability to analyze quickly, measure options against set criteria, and make decisions was joined in a productive tension to the group's rule of inclusiveness. Even the Branch Chief's inability to tolerate a collaborative process, tempered with his respect for the group members and their work, fed into the productive sessions at the end of the group's writing process. The result was a document that was accepted and used by almost all the negotiators in the Division and that required remarkably little rewriting at the end of eight months. (The reduction in the number of cost exhibits was the only major change.)

SUMMARY

This group, like all groups one is likely to encounter in the workplace, was less than ideal; however, the fact that group norms were established very explicitly in direct contrast to the wishes of the leader made it a valuable study in group process. Although few personal conflicts emerged and the initial stages of the group process worked well, the NET stagnated after these initial stages and might well have failed to generate a complete final product; almost certainly the group would not have met the deadline established. The stagnation was the result of the group's expansionary and inclusive norms, which had served the group well in earlier stages. The group was unable to change its own norms in order to meet the later demands of the problem-solving process. However, the group's leader and monitor, though not a skilled enough leader to actively direct the process, was skilled enough to know when his own behaviors were blocking the group and when these same behaviors would enhance the group effort.

IMPLICATIONS FOR GROUP FACILITATORS

This example suggests an option for the less-than-ideal group to facilitate the group's work. A leader/supervisor who knows what his or her own skills are could choose points of intervention in a group's work based on what the group needs at that particular time. In the case of the NET, the leader saw that his particular skills were needed at the final stages of the group's work; but it may well be that a supervisor whose group behaviors tend more toward the "expansionary" skills could get the group off to a good start by being involved in earlier stages, then stepping back to let the group finish the work. In either case, a leader must be aware of the group's norms and be alive to the possibilities for changes in norms resulting from changes in membership. When a leader steps in or introduces a new member into the group in order to provoke changes, she or he must consider the willingness of the group to change. A new, knowledgeable member will have more influence in the group if the group perceives itself to be failing, at least in some ways; the group may then welcome the newcomer as a catalyst for change [8]. However, even an obviously knowledgeable newcomer cannot solve the problem or complete the task unless other members are willing to let him or her do so [6]. Supporting the successful norms of the group is as essential to changing them as is introducing the change itself.

PART III:
New Implications for the Classroom

To help students improve writing and interpersonal skills and prepare for their next important, new discourse community, the workplace, technical communication teachers often require collaborative activities. Surveys have indicated that many professional writers engage in collaborative writing of some type (see, for example, Paul Anderson's review of survey results [1]). However, as Van Pelt and Gillam point out in their chapter, the goal of classroom collaborative assignments is student development. Each chapter in this Part demonstrates how the collaborative process helps students grow in both writing and thinking. In particular the authors discuss shared-document collaboration, or assignments in which groups of student writers share the authority for the final written product.

Each chapter offers case studies and classroom assignments and two include suggestions on how the computer can be used during the collaborative process ("courseware"). Technical communication instructors will find these examples easily adaptable to their classrooms. In general, the authors assume that writing is a social process and that instructors serve as "coaches," with minimal intervention in their students' collaborative groups. Rather than mastery of forms and documentation design standards, an appreciation and understanding of multiple perspectives becomes the goal of collaborative assignments. Two of the chapters link collaboration with the computer or telecommunications. The computer classroom offers an environment in which students negotiate ideas about audience, purpose, and content as well as write the document. The authors of all three chapters support their findings with close analyses of student-produced documents or with evidence from audiotapes, volume reports, questionnaires, or attitude surveys.

Elizabeth Tebeaux, in her "The Shared-Document Collaborative Case Response: Teaching and Research Implications in an In-House Teaching

Strategy," analyzes students' collaborative responses to cases based on company materials and real situations. Her cases challenge students to balance tact with honesty as they compose the required letters and memos. Placed in homogeneous groups by experience and by gender, Tebeaux's graduate students and employee workshop participants discovered how their perceptions might be a result of experience and background. Each group collaborated on a document, defended the document to the other groups, and then benefited from seeing the other groups' responses. Tebeaux concludes, "Recognizing the views of others involves recognizing the value of their perceptions. That is, there is not one way to design any response." Thus, the writers became "overtly aware of the importance of reader/purpose/tone considerations for all writing." Finally, Tebeaux speculates on how scholars might explore the effect of gender on writing and on how to bring both men and women to an "androgynous" style. Tebeaux provides the reader with both the case itself and several possible student responses, which should inspire technical communication teachers to develop similar assignments.

Ann Duin, Linda Jorn, and Mark DeBower, in their "Collaborative Writing—Courseware and Telecommunications," discuss ways that the computer or telecommunications can affect positively the collaborative process and the quality of student writing. Duin has won the "Best Curriculum Innovation in the Teaching of Writing" award from *Academic Computing* for courseware development. The courseware that Duin, Horn, and DeBower designed at the University of Minnesota can be adapted to many environments in which collaborators function. The authors offer extensive support for the link between computers and collaboration. They analyze the logs that student collaborators kept, audiotapes of group conversations, messages between collaborators on file servers, volume reports on how the telecommunications network was used by collaborators, student attitude surveys about the collaborative process, and reports from independent raters on the quality of writing produced by collaborators. As do Van Pelt and Gillam, Duin and her colleagues classify the type of "talk" between collaborators. Duin, Jorn and DeBower found six types of statements in the audiotapes they analyzed: strategy, audience, verification, content, technology, and off-task. To collaborate, group members spend a great deal of time questioning and clarifying individual views with their collaborators, a goal that Tebeaux affirms in her experience with collaborative case assignments.

The final chapter in this Part, William Van Pelt and Alice Gillam's "Peer Collaboration and the Computer-Assisted Classroom: Bridging the Gap between Academia and the Workplace," determines that although collaborative assignments may attempt to simulate to some extent workplace activities, the goals of each differ. While collaboration in industry affects productivity, classroom collaborative assignments must lead to intellectual and interpersonal student development. Van Pelt and Gillam first study industrial practices and then analyze classroom assignments. They link the computer with collaboration and assess the developmental writing and social issues that emerge during collaboration. By studying three groups of student collaborators, the authors trace student development in these new "interpretive communities." They found that students used four types of "talk" in the collaborative process to form group cohesiveness: procedural, substantive, writing, and social. Finally, because within the classroom individual intellectual development is a goal, Van Pelt and Gillam urge instructors to let students struggle through the collaborative process on their own, perhaps monitoring conflict only through student journals. Students must discover that written texts are "interpretive events constructed by communities of readers and writers out of their shared assumptions and on-going negotiations over discursive practices."

REFERENCE

1. P. Anderson, What Survey Research Tells Us about Writing at Work, in *Writing in Nonacademic Settings*, L. Odell and D. Goswami (eds.), The Guilford Press, New York, pp. 3–83, 1985.

CHAPTER 7

The Shared-Document Collaborative Case Response: Teaching and Research Implications of an In-House Teaching Strategy

ELIZABETH TEBEAUX

Perhaps the strongest mandate for including collaborative writing in technical communication courses comes from the workplace itself, where collaboration occurs extensively and in a variety of forms [1, p. 567; 2, pp. 50-51]. To prepare students for collaborative writing situations, a number of approaches have been developed to simulate those situations in work environments. Bruffee, for example, describes the use of the individual draft scrutinized during peer editing [3] ; Goldstein and Malone describe the development of the single document by a technical writing class [4]. Gebhardt suggests peer review for each stage of the writing process to help the writer perceive the effects of the message as it is developed [5]. Gere and Abbott stress peer responses to individually written drafts [6].

Ultimately, however, collaborative assignments have two goals: 1) to improve the individual's writing through increased sensitivity to group dynamics and sharpened awareness of how his/her writing is perceived; and 2) to prepare students for work environments where different forms of collaboration occur.

However, in an industrial short-course, collaborative writing, used with an approach often unavailable in academic classroom settings, can be particularly beneficial in helping employees fully realize the importance of planning every document based on perceptions of intended readers. As Aldrich concludes, lack of planning with readers and purpose clearly determined often causes employees to dread writing and to produce disorganized and ineffective writing [7]. However, throughout the short-course collaborative exercises, employees are constantly confronted with the perspectives and opinions of other group members who often have differing ideas that must be reconciled in developing the collaborative response. Thus, employees learn

1. the validity of defining the essential elements that control the design of any communication—audience, purpose, context, voice, and tone;
2. that perspective, based on a writer's background, job responsibilities, and job experience, determines how one writes and reads a document; and
3. that even collaboration does not negate the need for careful planning and analysis prior to writing.

A UNIQUE APPROACH TO COLLABORATIVE WRITING

During the past four years, I have conducted twelve workshops for employees who had difficulties organizing material and presenting it clearly. In negotiations with each company, I found that managers contracting these workshops wanted participants grouped by level. Their rationale was that participants would feel more comfortable if they were with others on their same organizational level. Thus, in six of the workshops, three each for two different companies, participants came from three organizational levels: entry-level managers, middle managers, and experienced managers. Entry-level managers had less than three years with the company. Middle managers had been with the company five to ten years, and experienced managers averaged about twelve to eighteen years of experience. In workshops for two additional companies, managers were either entry-level or mid-level.

In addition to two short report assignments developed individually, participants were given a final case problem that required a collaboratively written response. The case, like the topics for the individual report assignments, was developed from company materials and situations. In each short-course, each group of managers divided into groups of three or four did a thorough prewriting analysis of the case and then designed a shared response to the case. I evaluated each group's response to the case, made transparencies of these responses, and ultimately had all groups meet together during the final course session to allow participants to analyze how each group of managers had dealt with the case. The following case problem was developed for a collaborative assignment in a utility company's workshops. All managers, whatever their organizational level, would identify with the problem.

The Industrial Collaborative Writing
Problem — The Bill Finch Case

Robert Hansen, a friend who is a district-level supervisor in Network Design, has an opening for a second-line engineer. Hansen writes you a note saying that he has heard that Bill Finch, who is currently a customer services supervisor, might be a good person for him to consider. Hansen asks you to write him about your views of Finch. Hansen is a good friend of yours. You know that he has several employees who are not satisfactory, and you do not want to recommend anyone who might not work out. Finch was last evaluated ten months ago. In addition, you reflect, much has happened in the past six to eight months that might not be adequately reflected in Finch's personnel file. Some of these facts might affect Hansen's consideration of Finch. As you recall events of the past months, here is what you recall about Finch.

Bill Finch is a first-line foreman who is active in the community. You know about his activities because you both work in the same community improvement group. He is very active in the Lions Club and serves on the YMCA Board of Directors. On the job, his customer report rate has been above average but slightly below the objective. He has met his safety objectives and has good expense control—no problems there.

However, Finch has problems getting along with some of the people he works with. Five months ago he had a yelling match with another foreman. That scenario has been repeated at least four more times since then. The month before, he overstepped his bounds, telling another foreman's employee how to do a job when the regular foreman was available. Yet, Finch has the best crew on attendance: they met their objectives by a substantial margin. In spite of his problems with fellow workers, Finch is definitely a team player. For example, he was a key participant in family safety night last month.

While Finch is a good employee, you are concerned about his problems in getting along with people. About three weeks ago and then again last week, he failed to report to work without giving notice. However, you also know that Finch's wife Eloise has cancer, and her prognosis is not good. You overheard a good friend of Finch's comment that Eloise may not make it unless her response to treatment improves.

The problem you face is knowing just what to tell Hansen. You think the job change might help Finch. He certainly hasn't endeared himself to his current associates because of his unpredictable temper and moodiness. However, you think Hansen needs to know the facts. But after considering the situation, you decide you want Hansen to offer the position to Finch, as he needs the money and the job change. You now write a recommendation.

Rationale for the Shared-Document
Collaborative Assignments

Before discussing each group's response to this case, I need to explain the rationale for this particular application of the collaborative assignment. In industrial settings, this type of collaboration forces individuals to achieve what Wiener calls consensus [8]. Each person in the group shares responsibility for the final outcome produced by that group [9, pp. 84-85] . Once all responses from all groups are evaluated, each group of writers (as the following analysis of responses will illustrate) perceives the difficulty and the importance of analyzing the rhetorical elements that control the design of the message. In two previous studies, I found (as Aldrich did) that most employees who had writing problems lacked awareness of the perspective or needs of their readers [10, 11]. Not considering the reader in designing the reports discussed in these two studies resulted in writer-based messages that proved to be incomplete, visually dense, or poorly organized from readers' perspectives. While individual assignments allow employees to learn and practice principles of organization, format, and grammatical clarity to the development of a response, individually written responses are not nearly as effective as the shared-document assignment in helping employees understand that their way of perceiving a situation may not be the *only* way, that even those with whom they work often have differing (and perhaps more accurate) perceptions that must be accepted. In short, collaboration helps employees learn to plan messages by searching for and then harmonizing a variety of ideas and perspectives.

Collaborative Responses of Three Homogeneous Groups

Here's what happened when three homogeneous groups created shared-document collaborative responses to the Bill Finch case. There were two to four responses in each group, depending on the number of managers in a particular group. In evaluating the responses from each group, I was struck by the similarity of responses *within* each group but the dissimilarity of responses *among* the groups.

For example, one group of upper-level managers handled the Finch case in the following way, which was similar to the approaches used by the other group:

TO: Robert Hansen

FROM: xx xx xxxxxxxx

SUBJECT: Recommendation for Second-Line Engineering Opening

I concur with you on considering Bill Finch to fill the vacancy for the second-line engineering position. If selected, Bill will be beneficial to your organization, and the change will benefit him. You may be aware of the personal difficulty that Bill has had to deal with lately. If not, we can discuss it at your convenience.

Since ten months have passed since Bill's last performance evaluation, let me share his achievements with you:

Job Performance

Bill's customer report rate has been above average. He has met his safety objectives and has maintained good expense control.

Bill has the best crew on attendance. They met their objectives by a substantial margin. In addition, he is always a team player.

Community Activities

Bill and I are active in the same community group. He is making an outstanding contribution. He is also very active in the Lions Club and is currently serving on the YMCA Board of Directors.

Bob, I appreciate your considering Bill, as he is an excellent candidate. I am available at your convenience to discuss his record.

Middle-level managers took a more direct approach to the case, one that attempted to be positive (to recommend Finch) but to deal openly with Finch's problem. The following response was typical of their way of attacking the problem:

TO: Robert Hansen

FROM: x   x   xxxxxxx

SUBJECT: Position Vacancy—Second-line Engineer

I'm glad you called me about Bill Finch. Bill has been one of our better supervisors, and he is certainly deserving of promotion. I think you should take a good look at him in your search for a new manager. I'm forwarding a copy of his latest appraisals and other pertinent data, and I would also like to update you on his performance this year.

Bill has met or exceeded most of his objectives, most notably those of expense control, absence control, safety, and community relations. He is below objective on customer report rate, but above most of his peers. His community activities are particularly noteworthy for a first-line supervisor. He is on the YMCA Board, and he is active in both Lions Club and our community improvement group. He was instrumental in the success of our recent family safety night.

One area in which Bill still needs development is that of interpersonal skills. Although he's usually a team player, he is a strong-willed individual and has had occasional problems with some coordinates. You might sound him out on this when you interview him. He's had some family health problems recently, and I'm not sure how much that might have contributed to his problem. A change of environment and new challenges would probably have a positive impact on this situation.

I appreciate your consideration of one of my managers, and I hope this helps you in your decision. Bill has certainly helped my group in attaining our results and should be considered when opportunities like this arise. Call me if you need any further information, and I'll look forward to seeing you at the coordination meeting next week.

The responses of the entry-level managers showed the greatest difference in approach from those developed by the other two, more experienced groups of managers:

TO: Robert Hansen

FROM: xxxxxx   x   xxxxxxx

SUBJECT: Evaluation of Bill Finch

The following information summarizes my view of Bill Finch, Customer Service Supervisor–Residence I/M, since his last appraisal ten months ago:

Positive Points

- Customer Report Rate – above average but below objective
- Safety objectives met
- Good expense control
- Best crew on attendance – met objective by a wide margin
- Active in community – community improvement group, Lions Club, YMCA Board of Directors

Negative Points

- Problems coordinating with co-workers in past five months:
  Yelling match with another foreman five months ago—repeated at least four times since then.
  Last month told another foreman's employee how to do a job when the regular foreman was available.
- Two absences within the last three weeks without notice.

Summary

Bill's overall performance is good. I feel his problems in dealing with people and his absences may be related to his family problem. Bill's wife has cancer, and the outlook is not favorable.

Hopefully, this information will help you select a candidate for your position. Please let me know if I can provide further information.

The value of sharing these striking differences with all participants surfaced in the final joint session to discuss each group's responses. The combined groups viewed transparencies of each team's response. As each transparency was shown, the group that produced it explained the rationale for their response. The divergent approaches elicited intense debate, consensus defense by the group that wrote each response, quasi-admission of the legitimacy of divergent arguments, and, ultimately, a healthy respect for the importance of carefully considering audience perspective, purpose, tone, and organization in designing any message. This large-group session has one important advantage: participants who may be threatened by peer evaluation of their individual work are more comfortable in working with others whom they know and in helping analyze and defend the response they helped the group develop.

Group Evaluation of Responses

Even under scrutiny by others, less outspoken group members seemed comfortable responding as part of the consensus that produced a response. The difference in approach, which resulted from each group's level and experience in the organization, was soon evident.

Upper-level managers — These managers were appalled that specifics about Finch's problems would be discussed in the memos of the other two groups, as they believed that mentioning his wife's illness showed "poor taste" and simply was not done in "an organization like ours which is constantly subject to union arbitration." Several said that any mention of managerial problems would eliminate Finch from any further consideration. For that reason, since the case instructed the memo to be a recommendation, nothing negative should be mentioned. These managers agreed that Hansen needed some information about Finch's personal problem but only in a conversation when the impact of the information could be controlled. Upper-level managers were divided on whether the information could be shared by phone or whether it should be given in a face-to-face meeting with Hansen. These managers finally decided that the decision depended on the individual's telephone skills. One manager, however, continued to stress that the one-on-one discussion would be the only way to be sure that Finch would not be harmed by discussion of his home situation. As one senior manager observed, a memo like this one in Finch's file (pp. 128-129) could cost him this promotion and perhaps many more. As another observed, Finch does have a good record; no one knows at this point why he has shown a change in behavior; Hansen should be verbally informed of Finch's problems at home and some of his current work problems; and Hansen should be encouraged to make his own assessment. If he does not want Finch, that is his choice. Several managers commented on legal ramifications of specifically "airing" Finch's personal problems, particularly if he were to ever see the memo.

Middle-level managers — These individuals felt the need to be both positive and specific, although they finally agreed (during the group discussion) that the impact of explaining Finch's problems couldn't be controlled with any certainty in the memo. They intended to make the positive aspects of Finch's achievements overshadow his current problems, but they felt strongly committed to give the reader some indication that Finch had problems that might affect his job performance for Hansen. Middle-level managers were split over the effectiveness of paragraph one of the first memo (p. 127): some believed Hansen would be encouraged to call; others believed that Hansen might not call and either interview Finch or dismiss him as a viable job candidate entirely. They still felt that the memorandum should be "honest" and still provide a strong recommendation, although several upper-level managers stated adamantly that there is "no way to do that."

Entry-level managers — These managers were strongly censured by the other two groups of managers for their tactlessness. As one manager pointed out, "The first thing anyone sees [in the third memo, pp. 128-129] is 'negative points.' Once they read that, they won't read any of the 'positive points' with any credibility." Entry-level managers stated their belief in "giving the facts" whatever the result, even though the other managers convinced them that their approach would eliminate Finch from promotion. Several entry-level managers said they would feel more loyalty to a friend than to Finch, that Hansen deserved an honest, detailed assessment. As one newly promoted upper-level manager with fifteen years' experience observed, "that kind of attitude is what keeps people down. As a manager, you are supposed to help and develop people who work for you. One bad letter in a file can ruin someone. Besides, what if you are wrong about Finch? None of us ever knows *all* the facts about anyone. The effective manager is the one who assesses the situation as carefully as possible and tries to act as constructively as possible in terms of both the individual and the company." An entry-level manager stated that not documenting Finch's problems was what caused people to get jobs they shouldn't have. One senior manager then snapped that the statement about Finch's dilemma in paragraph one of the first memo (p. 127) "should tell you clearly that you should call me" and that "the ability to discriminate is what a senior manager must have."

The comments of each group became particularly relevant when entry-level managers became aware that they would likely be writing memos like the one called for in the case. They seemed to realize that decisions about what to say, how to say it, and what interpretations could be attached to specific statements, and the long-range results of these interpretations were more complex than they had realized. Many middle-level managers had already had to write similar memos and seemed to become more aware of the long-term, not just the immediate implications of what they chose to say. Writers in both groups openly admitted that they often wrote personnel reports without considering how their comments could be construed. Upper-level managers frequently had to write reports recommending middle-level managers for promotions. Several upper-level managers asked: "How would any of *you* guys like it if I wrote a memo, supposedly recommending *you* for a promotion, and discussed all *your* family problems? Put yourself in Finch's shoes. How does it feel?"

Results of the Combined Group Analysis

A number of realizations emerged from this confrontation of teams:

1. One's managerial experience determines one's perspective on a given problem. This statement reflects Odell's suggestion that "In judging the appropriateness of choices that appear in their writing, writers . . . relied on their awareness of attitudes and prior experiences that are shared by others in their organization" [12, p. 251].
2. One's job level determines perception of information given and information received. Or as Odell also observed, writers' perceptions of their audience reflected not only their knowledge of the intended reader, but also their understanding of their own job and the attitudes of the particular office in which they work.
3. Accurately defining the views and perceptions of intended readers is extremely tricky. Each group, prior to the final session, had been thoroughly comfortable with the response generated.

However, the most important result involved a change in perspective, a new respect for the necessity of designing messages. Participants seemed to achieve a new awareness that, as Drucker would say, "communication is perception [13, p. 483]:"

4. Recognizing the views of others involves recognizing the value of their perceptions. That is, there is no one way to design any response.

More experienced managers recognized that they were more sensitive to the metadiscourse of messages—the implied meaning as it arises from careful structure of sentences and use of hedging, indirect phrases—than entry-level managers were. Experienced managers had learned, from their job experience and from early mistakes, the value of understatement and tact, while managers with little experience tended to overestimate the importance of facts while undervaluing tact. Middle-level managers also saw the problems in trying to be specific and positive at the same time, that this combination of purposes could be construed as an attempt to "whitewash" Finch's problems rather than as a *sincere* attempt to recommend him while cautiously giving specific information. Several entry-level managers admitted that writing at work was very different from writing at school and commented that academic writing was of little value in dealing with job-related writing. They pointed out that in school one is expected to provide detailed presentation. On the job, conciseness and political ramifications were the major issues. Mirroring conclusions by Paradis et al., they were finding the adjustment difficult and frustrating [14, p. 300–303].

In summary, the three groups, in their analysis of the collaborative group efforts, discovered the following characteristics about themselves and each other:

Upper-level managers were
- more sensitive to potential problems in any response;
- less sensitive to tone. They tended to choose content carefully to present it "objectively;" and
- less inclined to write too much. Other managers viewed upper-level managers as not detailed enough. Upper-level managers have so much to read that they expressed definite distaste for lengthy documents.

Middle-level managers were
- more concerned about whether their message would be read as they intended it. (Upper-level managers just assumed what they wrote would be understood.);
- less aware of the potential hazards of content; and
- aware of the importance of tone but not always consistent in knowing how they sounded.

Lower-level managers were
- more concerned with the "facts of the case;"
- inclined to write "too much" in the opinion of both higher groups of managers;
- insensitive to tone; felt that facts and details should be the focus of writing;
- not aware of multiple purposes or political hazards of the message; and
- not exactly sure—at the beginning of the analysis—why what we were doing was important.

Benefits of the Shared-Document Assignment

In the anonymous written evaluation that each participant was asked to complete at the conclusion of the workshop, employees in all eight workshops (in which this type of collaborative assignment was used) rated the workshop as "excellent" or "very good." Over 70 percent gave the workshop an "excellent" rating, while 81 percent of the employees checked the collaborative exercise as the "best" assignment. In the "Comments" portion of the evaluation sheet, 42 percent of the employees stated that they benefited from understanding how employees on other levels viewed difficult management decisions, such as the Bill Finch recommendation, and 66 percent noted that a changed perception toward the writing process was the most useful benefit they derived from comparing and analyzing each group's shared document. Finally, 83 percent said that what they had learned in the workshop made them feel more comfortable about writing and knowing how to go about determining content.

Several managers commented that the demands of one's own job make analysis of readers difficult and in many cases impossible. Others agreed that the hazards of writing are not considered frequently enough during the writing process. However, participants recognized that discussing their collaboratively produced responses had helped them come to terms with their own attitudes while being able to listen to the views of others. Employees, even more than students, tend to believe that there is one way to write something. However, the need for editing to achieve conciseness and clarity was never questioned, as most managers realized that they could not read all of everything that crossed their desks. Yet, as Odell et al. discovered, employees, like these upper-level managers, do learn the importance of analyzing readers, purpose of the message, and even tone without consciously knowing that they are performing rhetorical analysis [15, p. 27–36]. The greatest value of the shared response generated within homogeneous employee groups may lie in helping all writers become overtly aware of the importance of reader/purpose/tone considerations for all writing.

HOMOGENEOUS COLLABORATIVE GROUPS— IMPLICATIONS FOR BUSINESS AND TECHNICAL WRITING INSTRUCTION

As a result of this experience, in graduate-level professional writing classes, I have grouped individuals for shared document assignments in two ways: by undergraduate major and then by work experience. The results have been as follows:

First grouping—Collaborative assignments developed by teams grouped by specific majors showed relatively little difference among the responses. That is, M.B.A.'s with liberal arts backgrounds did not handle the response differently from students with undergraduate engineering or science backgrounds.

Second grouping—Collaborative assignments that were developed by teams grouped according to their work experience showed greater divergence among responses. M.B.A.'s with significant work experience that involved managerial responsibility, as compared to students with little work experience, showed the following characteristics:

... greater awareness of the importance of analyzing audience, purpose, tone, and voice;

... greater awareness of differences between academic writing and writing at work;

... greater interest in learning to deal with tone and voice; and

... better knowledge of how to deal with the problem in the case. Greater awareness of the need to incorporate CYA strategies in many documents.

In short, job experience requiring an employee to work closely with others in supervision or collaborative discussion helps employees develop skills in analyzing audiences and then in developing messages that will achieve their purpose in the business context.

USING COLLABORATIVE ASSIGNMENTS TO ASSESS GENDER DIFFERENCES

If job experience requiring interpersonal interaction can modify an employee's writing, will similar job experience also affect gender-based differences in collaboratively developed written business communications? Graduate-level professional writing classes may allow the creation of collaborative homogeneous groups—separated by both work experience *and* by gender—to study two additional research issues: 1) to what extent are gender-based writing characteristics evident in written business communications? and 2) do gender and work experience affect the quality of written case responses or how well the response achieves its purpose with the intended audience(s)?

Psychological gender studies reveal that females and males have distinctively different interpersonal skills. For example, Chodorow believes that girls have an inherent empathy, while boys display a separateness in their dealing with others [16]. Similarly, Gilligan suggests that one characteristic of femininity is an ethic of caring, while an ethic of objectivity is characteristic of masculinity [17]. Lay, in summarizing research on psychological gender differences, supports these observations and suggests that the qualities of concern, empathy, sensitivity, and receptiveness—female characteristics—are qualities essential to male and female collaborators [18].

A substantial body of research suggests that written and oral communications of men and women reflect these differences in both style and content. Gender-based communication research indicates that men in speaking contexts use more dynamic language [19, p. 250-251], more quantitative and objective adjectives [20, p. 137], more aggressive verbs [21, 22], while women use more descriptive, aesthetic, emotive language [23, 24], intensifiers [25], interpretive adjectives, hedges, tag questions, disclaimers, intensifiers, requests (rather than demands) [26-28].

However, only two reported studies have sought to determine if differences exist in the written business communications of women and men. Neither Sterkel [29] nor Smeltzer and Werbel [30] found differences in the styles or quality of business communication students. However, neither study specifically considered the role that work experience might play in highlighting or diminishing gender-based writing differences, and Smeltzer's group, nearing completion of the M.B.A., had already completed a graduate-level course in business communication [30, p. 43]. Sterkel's group was comprised of junior or senior business students, but their work background apparently was not assessed [29, p. 34]. My preliminary work with a homogeneous group of graduate business communication students suggests that both work experience and gender may have significant bearing on the quality—meaning that the message achieves the writer's purposes—of written business communication case responses.

For example, in a written communication course I teach for graduate students in the M.S. Tax program, the course, launched in 1984, consists of usually half male and half female students. While all have a bachelor's degree in accounting, their experience levels vary and usually fall into four categories: women who have accounting or business experience (2-4 years); men who have accounting or business experience (2-8 years); men with no business experience; and women with no business experience. Students with no work experience usually enter the M.S. Taxation program immediately after completing the B.S. in Accounting. Some of these students have some kind of summer work experience, but most of their work experience does not require them to work closely with the public, with clients, or with other employees in problem-solving situations. However, students with experience have worked for an accounting firm for several years or been involved with corporate accounting for a major company. In seven shared-document assignments given to four different tax communication classes in which students were grouped according to these classifications, I have found differences in responses that both support and refute studies on gender differences.

The following case was developed for the 1987 tax students in their shared-document groups. The situation was one with which everyone in the class was familiar, as the accounting department sponsors an annual tax conference which is planned each year by the tax students. Again, the challenge was how to achieve the purpose most effectively.

The Academic Collaborative Case—
The Annual Tax Conference

Mr. Albert McGee is a senior partner with one of the major private accounting firms in your part of the state. Because of his reputation in accounting and in community leadership, he has been a speaker at the annual tax conference attended by accountants from firms throughout the area. The conference is sponsored by Hayward State University, which has a well-known accounting department and tax program. You are a CPA who teaches part-time at Hayward State, but you work for McGill, Smith, and King, located near the Hayward campus. This year, you have been asked by the accounting department to chair the arrangements committee for the tax conference.

McGee helped establish the tax conference. In addition, his firm has funded two endowed chairs for the accounting department at Hayward. McGee has been a speaker every year during the nine years the tax conference has been held at Hayward. For five years, he was the star speaker. The problem, however, is that he had a stroke three years ago, and his last two presentations have been dreadful. While you admire his determination to recover, your planning committee has decided that he simply cannot be asked to deliver a presentation again this year. Several accountants who attend annually to receive CPU credit have already called to ask members of the committee "what they plan to do about McGee." You are pretty sure that if McGee is scheduled, out of respect for his past contributions, no one will attend his presentation, as there are two presentations scheduled concurrently at each two-hour session. Everyone on the committee is concerned that having McGee face an empty room would be worse than not inviting him.

The problem is difficult because Joey Reynolds, who is an expert on estate tax—McGee's area—has already been invited to speak on new issues in estate tax planning. Reynolds is brilliant, and his presence will increase the number of people who will attend the conference.

You and the committee agree that the conference needs a rule that prevents speakers from giving more than three consecutive presentations. The committee votes unanimously to include this statement in the conference guidelines. However, you still have the problem of what to do about McGee. Reynolds has been invited, so you decide to write McGee a letter informing him about the new guideline. The conference is three months away, but you have the announcement which includes the program ready. It will be printed and mailed in two weeks to all the area accountants.

You must inform McGee today that he will not be a speaker. You do not want him to receive the letter about the time the conference announcement arrives at his office. After you meet with the planning committee, you return to your office. Your secretary hands you a letter from McGee that contains his speech outline—which is incomprehensible—and a letter telling you that he is looking forward to the conference.

Evaluation of Response

Evaluation of the responses showed that the most striking difference occurred between men and women with corporate business or accounting experience and those who were recent college graduates having only summer work experience. Writers with experience were more tactful than nonexperienced writers, who were blunt and objective in presenting "the facts." These differences mirrored those described in the in-house collaborative assignment. In the 1987 class, the four groups produced the following letters. Interestingly enough, the shared-document groups of the 1988 class, dealing with the same case, produced similar responses. The first letter was written by women who had several years' experience in accounting or business:

Dear Mr. McGee:

In response to your letter regarding this year's tax conference, I must inform you of a new rule which the arrangements committee has included in our conference guidelines. Basically, the new rule prevents speakers from giving more than three consecutive presentations at our annual tax conference. This means that you and several other of our regular speakers get to take this year off and can just enjoy listening to the other speakers.

The committee hopes the new rule will serve two basic purposes. First, we hope to enhance the value of our conference by exposing the audience to different speakers' viewpoints each year. The audience, as well as prior years' speakers, will get the opportunity to compare and contrast these different viewpoints and possibly combine strategies that will help them in their practice.

Second, the committee would like to offer many different people the opportunity to speak at our annual conference. We feel that the speaking experience helps to improve the quality and performance of each presenter in his/her own careers. After all, you yourself are a prime example of the benefits that can be achieved by becoming a good speaker.

As you can see, the committee is making every effort to enhance the benefits of the tax conference to all involved. Founders of the conference, such as you, have contributed greatly to the past success of the conference. Now it is our turn to do our part.

I look forward to seeing you again at this year's conference. Thank you for your support of our program.

This collaborative assignment was given in both classes after general principles of correspondence had been presented. Students had already written two individually composed letter responses to tax case problems. Tone and structure of letters had been presented, and students had critiqued each other's work via transparencies during class. As the above letter indicates, this group of women writers designed an effective way to state negative news, justify it in a positive way, and state the key issues in the opening paragraph. This approach has the value of not misleading Mr. McGee, who might not read the remainder of the letter and miss the main point if it were buried in later paragraphs. As the case states, he is not expecting a letter informing him that he will not be speaking this year. Students in each collaborative group agreed that the purpose of the letter was to inform McGee that he would not be speaking and to do so in such a way that his financial support would be maintained and good relationships between him and the tax faculty would not be damaged.

The next letter, written by experienced male writers, does a tactful job of telling McGee he won't be speaking, but the bad news does not occur until the middle of paragraph three, and it is extremely blunt. The group made no effort to explain the new rule. However, this group thought that making McGee the guest of honor would be an appropriate strategy to soften the news and keep McGee's good will:

> Dear Mr. McGee:
>
> As you are aware, our annual tax conference is only three months away. We would like it very much if you would accept our invitation to be our guest of honor at the luncheon.
>
> You helped establish the tax conference, and your firm has funded two endowed chairs for our accounting department. You have been a speaker for the past nine years and our star speaker for the past five years. We cannot think of a better candidate to be our guest of honor.
>
> Since you have been such an active participant in our conference each year, I would like to inform you of the recent changes which have taken place. The arrangements committee has added a new rule to the conference guidelines. The new rule will prevent speakers from giving more than three consecutive presentations. Since you have been our star speaker for more than three years, I regret to inform you that you cannot be asked to give a presentation at this year's conference. We would like to be able to invite you to continue in your capacity as speaker, but we feel we must follow the new guidelines. I hope you understand our predicament.
>
> We would greatly appreciate, however, your acceptance as our guest of honor for the luncheon. If you have any questions, please call me at xxx xxx-xxxx. I look forward to seeing you again.

The third letter was developed by women writers who had little or no business experience. Some had summer bookkeeping experience or clerical experience, but they had entered the tax program immediately after completing the B.A. in accounting. This group took an attack similar to the one taken by the men in the previous letter; but like the first letter, the tone reflects a concern for McGee's feelings and offers to make him guest of honor. Like the first letter, it does an excellent job of handling bad news in a positive way, but like the second letter, the actual fact does not occur in the opening paragraph.

Dear Mr. McGee:

As Chairman of the Hayward College 1987 tax conference planning committee, I very much look forward to meeting you at the conference. Although I have not met you, I would still like to thank you for your genuine interest in our tax conference. It takes a special individual to contribute so much of his own time and financial contribution to a program he believes in.

I am writing to you today, Mr. McGee, to tell you about some new guidelines that my committee enacted to improve the quality of the tax conference—a conference that could not have been established without your help. The new guidelines prevent speakers from giving more than three consecutive presentations. This guideline is meant to ensure variety in speakers as well as topics.

Because of this new guideline and on behalf of the planning committee, I respectfully ask that, in lieu of having you speak at the conference, you be our guest of honor at the reception on the eve of the tax conference. The committee unanimously decided that it is past time to thank you officially for all your contributions to the College of Business. We would be honored to have you accept this token of our appreciation. A more formal invitation will be forthcoming.

It may be of interest to you that we have invited Mr. Joey Reynolds to speak on new issues in estate tax planning. He is very adept in his subject, and I think you will be pleased with his presentation.

I will be in touch with you soon about the reception.

The least effective of the four collaborative responses was written by men with no business experience. What work experience these male students had did not require them to deal with people. Their response states the bad news bluntly in the opening paragraph, and like inexperienced writers dealing with the Bill Finch case, this group of writers struggles with their need to "be honest" and the need to be tactful. In fact, their efforts at being tactful sounded contrived to the other three groups.

Dear Mr. McGee:

We just came from a tax conference planning meeting in which we instigated a rule that no one may speak at more than three consecutive conferences. We felt this rule was needed so that we can give more people the opportunity to speak and so we can expose the audience to different speakers and viewpoints. Also, it will allow veteran speakers like you to relax and enjoy the conference.

We intended this rule to be a positive change for all involved, but I feel you may view this as an attempt to cut you out of the conference. On the contrary, we look forward to having you as our honored guest. You will, of course, continue to be placed in the seat of honor at the banquet. I also look forward to any ideas or help on the actual planning of the conference.

I look forward to discussing plans for our conference with you and will call you early next week to touch base with you. If anything comes up before then, please give me a call.

Considering Gender Differences

In examining responses to the four shared-document assignments developed for students enrolled in the tax communication course, I have found that differences in messages can be observed between students having significant work experience and those having little or no work experience. But the responses also show that women tax students respond more tactfully than men do in case problems requiring tact, but men and women with work experience and even women with little work experience developed a more effective (tactful) reply than men with no work experience. Again, effectiveness means a response that achieves both the human objective and the business objective, to use Bowman and Branshaw's phrase [31, p. 40]. And, the greater the need for tact, the better experienced men and women handle the case and the worse inexperienced men respond. Females with little significant work experience usually give unsatisfactory responses to difficult cases—usually omitting important ideas—but show dexterity in phrasing difficult ideas. In addition, while these differences often surface in a comparison of individually written responses, they are accentuated in collaborative assignments where men and women are separately grouped, according to their work experience, to design responses. The difference, however, is not one that is easily measured—such as use of negative words, passive voice, "you" attitude, and other specific style characteristics examined by Sterkel [29, pp. 23-24] and Smeltzer and Werbel [30, pp. 44-45]. Many times, the responses by men and women in the collaborative groups have been equally effective, as determined by the entire class evaluating the fifth and sixth memos (pp. 138, 139); but men, even the experienced writers in this population of students, tend to be more direct than women in delivering the message. Women seem more aware of phrases and even entire approaches that produce negative tone and attitude toward the reader.

These tendencies have surfaced when the entire class critiqued each group's response. As in the employee collaborative group assignment, the student group that wrote the response had the opportunity to explain the rationale for the approach they used. During the discussion of each group's response, women, as a whole, were more concerned that McGee's feelings be spared as much as possible, while men sought to be tactful but *specific* in telling him that he would definitely *not* be speaking. These findings support observations by Gilligan [17], Maltz and Borker [32, pp. 196-216], Thorne and Henley [33], Baird [34], and Eakins and Eakins [27, pp. 6, 48] that women are more caring and that this quality surfaces in their language.

However, these differences totally evaporate in the purely technical assignments written by these tax students—i.e., students writing either individually or in shared-document groups to produce research memoranda for the IRS on tax problems or internal reports for senior tax managers or partners. In these assignments, where the purpose is to deliver specific, accurate, factual information and where personal feelings, politeness, and tact are not factors, no gender differences surface. In these assignments, knowledge of the IRS Code and the ability to write sentences that exemplify grammatical clarity and correct usage are the main requirements for effectiveness. Thus, male students who are tactless in writing case responses may perform as well as male or female students with extensive accounting backgrounds. This contrast suggests a number of possible questions needing further research:

1. Does significant job experience help develop androgynous or gender-free language, used by both men and women, so that both display a verbal flexibility that neutralizes gender differences shown to exist in non-business contexts? Do men with extensive business experience continue to show more aggressive, instrumental approaches than women, who seem to prefer phrases that are more empathetic and less functional than those used by experienced men?

2. Do the suggested relationships between work experience and gender discovered in tax students apply to other groups? If M.B.A. students were categorized by gender and work experience, would their responses to business communication cases reflect the similar differences in the quality of the communication—the appropriateness of the written response to the case problem?

3. If experienced and gender-based differences exist in business communications requiring tact, do these same differences exist in technical assignments, such as marketing research reports, proposals, specifications, economic analyses, and work procedures? If these differences exist, do they affect the quality of these documents, their ability to achieve the writer's purpose?

4. Can case studies requiring extraordinary tact be used to help students develop different styles to deal with difficult issues? Can this practice of adjusting tone and style to audience help students recognize bluntness when it is inappropriate or excessive politeness when a more direct approach is required by a case problem?

Studies by Talley and Richmond [35] and Key [36] suggest that androgynous language is prevalent. As Talley and Richmond state, "Many individuals might be 'androgynous,' both male and female, both independent and compliant, both dominant or submissive. This concept refers to the tendency for individuals to display sex-role flexibility across situations without regard for stereotyping behaviors as more appropriate for one sex or another" [35, p. 328]. Key further recognized that "the linguistic behavior of men and women in the office or at work might be almost identical, but the same persons will exhibit other linguistic patterns in other situations or in other roles" [36, p. 16].

Knowing the gender-based writing difficulties that students may have, given their work background, can become a powerful tool in guiding students and helping them overcome strict roles and language characteristics that hinder their inability to deal with sensitive written communication problems. In each class of tax students, I have noted that continued practice on cases requiring sensitive content presentation increases all students' ability to write in a variety of voices necessary to solve the many technical and human communication problems a tax accountant faces.

EVALUATION OF HOMOGENEOUS
COLLABORATIVE GROUP ASSIGNMENTS

Existing discussion of academic collaborative assignments has already shown how collaboration can help students understand how their writing is perceived by others, not just by a teacher. However, the use of homogeneous collaborative groups has been generally underused because most writing classes, with their varied student populations, do not allow grouping by factors such as job experience. However, even in classes where students have similar amounts of work experience, collaborative assignments can be assigned by gender and then by random groupings or by equal numbers of men and women in each group to allow investigation and then comparison of gender-based communication differences that may surface.

However, collaborative case responses developed by homogeneous groups seem to accentuate the defining qualities and characteristics of those groups. In turn, by comparing their response to the responses of other homogeneous groups, each group can be led to understand its own perspective and the importance of defining reader/writer perspective in designing all writing. At the same time, collaborative responses designed by all-men and all-women groups help participants as well as teachers spot writing differences, difficulties, and perspectives that may be gender specific and ultimately detrimental to a student's performance in the workplace. Research in the 1960s and 1970s suggested that "women's language"—characterized by empathy, submissiveness, and meekness— led to devaluation of women's management skills and ineffectiveness in dealing with people [37-39]. However, more recent studies recognize the value of feminine communication styles in enhancing the credibility of the ideas presented [40, 41]. Both Bem and Bradley see the importance of an individual being able to choose roles. As Bem states, "men and women should be encouraged to be both instrumental and expressive, both assertive and yielding, both masculine and feminine—depending on the situation and the appropriateness of these various behaviors" [42, p. 639]. Bradley is even more specific [43, p. 90]:

What needs to be reassessed, then, is the discriminating way that this culture views men and women, leading to the evaluation of communicative acts performed by the former and the devaluation of those performed by the latter. The person who chooses a tactful, sensitive method for expressing a controversial idea should probably be praised, whether that person is a man or a women. Only when the appropriateness of communicative acts are judged by the extent to which they are fitting in the situation in which they are used will improved human interaction and decision-making processes likely occur.

McEdwards remarks that the qualities of women's language are often used by both men and women in situations in which powerless is inherent [40]. As Lay states, by "allowing a fuller range of behavior to males, by incorporating within our definitions of masculinity such traits as nurturing, and by encouraging ultimately androgynous identity, we should allow men as well as women to be comfortable and effective collaborators.... The effective collaborator, then, would have full access to the interpersonal tools we too often label masculine or feminine" [18, p. 12].

Courses in technical and business communication that make extensive use of personnel problem cases developed individually or collaboratively are providing students the education they need to prepare them for a work context where effective collaboration—networking [44], small-group problem solving [45] —will be critical to the organization's profit and excellence [46]. Thus, continued research in collaborative pedagogical approaches is essential to make these methods even more effective by enhancing our understanding of gender-based communication differences and the role that education and job experience contribute to diminishing these differences.

REFERENCES

1. L. Faigley and T. Miller, What We Learn from Writing on the Job, *College English, 44*:6, pp. 557–569, 1982.
2. P. V. Anderson, What Survey Research Tells Us about Writing at Work, in *Writing in Nonacademic Settings,* Lee Odell and Dixie Goswami (eds.), The Guilford Press, New York, and London, England, pp. 3–83, 1985.
3. K. A. Bruffee, Collaborative Learning and the "Conversation of Mankind," *College English, 46*:6, pp. 635–653, 1984.
4. J. R. Goldstein and E. L. Malone, Journals on Interpersonal and Group Communications: Facilitating Technical Project Groups, *Journal of Technical Writing and Communication, 14*:2, pp. 113–131, 1984.
5. R. Gebhardt, Teamwork and Feedback: Broadening the Base of Collaborative Writing, *College English, 42*:1, pp. 69–74, 1980.
6. A. R. Gere and R. D. Abbott, Talking about Writing: The Language of Writing Groups, *Research in the Teaching of English, 19*:3, pp. 362–381, 1985.
7. P. G. Aldrich, Adult Writers: Some Reasons for Ineffective Writing on the Job, *CCC, 33*:3, pp. 284–277, 1982.
8. H. Wiener, Collaborative Learning in the Classroom: A Guide to Evaluation, *College English, 48*:1, pp. 52–61, 1986.
9. N. Allen, D. Atkinson, M. Morgan, T. Moore, and C. Snow, What Experienced Collaborators Say about Collaborative Writing, *Journal of Business and Technical Communication, 1*:2, pp. 71–90, 1987.

10. E. Tebeaux, The Trouble with Employees' Writing May Be Freshman English, *Teaching English in the Two Year College, 15*:1, pp. 9-19, 1988.
11. _____, Writing in Academe; Writing at Work: Using Visual Rhetoric to Bridge the Gap, *The Journal of Teaching Writing, 7*:2, pp. 215-236, 1988.
12. L. Odell, Beyond the Text: Relations Between Writing and Social Context, in *Writing in Nonacademic Settings*, Lee Odell and Dixie Goswami (eds.), The Guilford Press, New York and London, England, pp. 249-280, 1985.
13. P. F. Drucker, *Management: Tasks, Responsibilities, Practices*, Harper and Row, New York, 1974.
14. J. Paradis, D. Dobrin, and R. Miller, Writing at Exxon ITD: Notes on the Writing Environment of an R&D Organization, in *Writing in Nonacademic Settings*, Lee Odell and Dixie Goswami (eds.), The Guilford Press, New York and London, England, pp. 281-307, 1985.
15. L. Odell, D. Goswami, and D. Quick, Studying Writing in Nonacademic Settings, in *New Essays in Technical and Scientific Communication: Research, Theory, and Practice*, Lee Odell and Dixie Goswami (eds.), Baywood Publishing Company, Inc., Amityville, New York, pp. 17-40, 1983.
16. N. Chodorow, *The Reproduction of Mothering: Psychoanalysis and the Sociology of Gender*, University of California Press, Berkeley, 1978.
17. C. Gilligan, *In a Different Voice: Psychological Theory and Women's Development*, Harvard University Press, Cambridge, Massachusetts, 1982.
18. M. M. Lay, Interpersonal Conflict in Collaborative Writing: What We Can Learn from Gender Studies, *Journal of Business and Technical Communication, 3*:2, pp. 5-28, 1989.
19. A. Mulac and T. L. Lundell, An Empirical Test of the Gender-Linked Language Effect in a Public Speaking Setting, *Language and Speech, 25*:3, pp. 243-256, 1982.
20. M. M. Wood, The Influence of Sex and Knowledge on Communication Effectiveness in Spontaneous Speech, *Word, 22*, pp. 117-137, 1966.
21. E. F. Sause, Computer Content Analysis of Sex Differences in the Language of Children, *Journal of Psycholinguistic Research, 5*, pp. 311-324, 1900.
22. R. Westmoreland, D. P. Starr, K. Shelton, and Y. Pasadeos, News Writing Styles of Male and Female Students, *Journalism Quarterly, 54*, pp. 599-601, 1977.
23. G. C. Gleser, L. A. Gottschalk, and W. John, The Relationship of Sex and Intelligence to Choice of Words: A Normative Study of Verbal Behavior, *Journal of Clinical Psychology, 15*, pp. 182-191, 1959.
24. A. Mulac and M. J. Rudd, Effects of Selected American Regional Dialects upon Regional Audience Members, *Communication Monographs, 44*, pp. 185-196, 1977.
25. N. L. Colwell and T. I. Sztaba, Organizational Genderlect: The Problem of Two Different Languages, *Business Quarterly, 24*, pp. 64-66, 1986.
26. R. Lakoff, Women's Language, *Language and Style, 10*, pp. 222-247, 1977.
27. B. W. Eakins and R. G. Eakins, *Sex Differences in Human Communications*, Houghton Mifflin Company, Boston, 1978.
28. F. Crosby and L. Nyquist, The Female Register: An Empirical Study of Lakoff's Hypothesis, *Language and Society, 6*, pp. 313-322, 1977.

29. K. S. Sterkel, The Relationship Between Gender and Writing Style in Business Communication, *The Journal of Business Communication, 25*:4, pp. 17-38, 1988.
30. L. Smeltzer and J. D. Werbel, Gender Differences in Managerial Communication: Fact or Folk-Linguistics? *The Journal of Business Communication, 23*:2, pp. 41-50, 1986.
31. J. P. Bowman and B. P. Branshaw, *Effective Business Correspondence,* Harper and Row, New York, 1979.
32. D. N. Maltz and R. A. Borker, A Cultural Approach to Male-Female Communication, in *Language and Social Interaction,* John J. Gumperz (ed.), Harvard University Press, Cambridge, Massachusetts, 1982.
33. B. Thorne and N. Henley, Difference and Dominance: An Overview of Language, Gender, and Society, in *Language and Sex: Difference and Dominance,* B. Thorne and N. Henley (eds.), Newbury House, Rowley, Massachusetts, pp. 5-42, 1975.
34. J. E. Baird, Jr., Sex Differences in Group Communication: A Review of Relevant Research, *Quarterly Journal of Speech, 62,* pp. 179-192, 1976.
35. M. A. Talley and V. P. Richmond, The Relationship Between Psychological Gender Orientation and Communication Styles, *Human Communication Research, 6*:4, pp. 327-339, 1980.
36. M. R. Key, Linguistic Behavior of Male and Female, *Linguistics, 88,* pp. 15-31, 1972.
37. C. Kramer, Women's Speech, Separate But Equal? *Quarterly Journal of Speech, 60,* pp. 14-24, 1970.
38. K. A. Foss and S. K. Foss, The Status of Research on Women and Communication, *Communication Quarterly, 31,* pp. 195-203, 1983.
39. F. Johnson, Political and Pedagogical Implications of Attitudes Towards Language, *Communication Quarterly, 31*:2, pp. 133-138, 1983.
40. M. G. McEdwards, Women's Language, A Positive View, *English Journal, 74,* pp. 40-43, 1985.
41. C. L. Berryman-Fink and J. R. Wilcox, A Multivariate Investigation of Perceptual Attributions Concerning Gender Appropriateness in Language, *Sex Roles, 9,* pp. 663-681, 1983.
42. S. L. Bem, Sex-Role Adaptability: One Consequence of Psychological Androgyny, *Journal of Personality and Social Psychology, 31,* pp. 634-643, 1975.
43. P. H. Bradley, The Folk-Linguistics of Women's Speech: An Empirical Examination, *Communication Monographs, 48,* pp. 73-90, 1900.
44. V. H. Hine, Networks in a Global Society, *The Futurist,* pp. 11-13, June 1984.
45. S. Fraker, High-Speed Management for the High-Tech Age, *Fortune, 5,* pp. 62-68, March 1984.
46. T. Peters and R. Waterman, *In Search of Excellence,* Harper and Row, New York, 1982.

CHAPTER 8

Collaborative Writing— Courseware and Telecommunications

ANN HILL DUIN
LINDA A. JORN
MARK S. DeBOWER

Writing is essentially a social and collaborative process. We write and share writing in order to engage other people's attention and interest, to tap their expertise, and to augment our understanding of projects and organizations [1]. Theorists speak of the power of using computer technology to enhance this collaborative writing process. Lyman states that "technology must be a catalyst for rethinking teaching and learning, redesigning the pedagogic strategies of the teacher and broadening the contexts within which learning can occur" [2, p. 18], and Daiute concludes that "collaborative writing should be motivated by the goal of setting writing in communication contexts" [3, p. 49].

The main goals of this project were to help students learn how to collaborate or coauthor nonacademic documents, to help students cooperate and give feedback on other students' documents, and to help students use telecommunications as a means to collaborate and give and receive feedback. To achieve these goals, we designed courseware (educational software) for collaborative writing and integrated this courseware with telecommunications technology.

We designed the courseware because our research suggested that workplace writing involves a great deal of collaboration. Specifically, professional collaborative writing tasks include a range of activities involving a supervisor's assignment of a document to be written by a staff member and later edited by the supervisor, group planning of a document that is drafted and revised individually, individual planning and drafting of a document that is revised collaboratively, peer reviews of co-workers' drafts, and coauthoring of documents [4-11].

We then integrated the courseware with telecommunications because our research suggested that telecommunications is an important skill for students to possess [12, 13]. From a survey of recent business graduates working in Fortune 500 corporations, Bednar and Olney found that the most frequently used communication forms are memos, computer networks, and information reports, and that computer networks are used more than letters or analytical reports. They also found the three most serious communication problems among graduates to be poor listening, lack of conciseness, and poor feedback [14].

We also surveyed courses at our university and found that despite being a common form of composing in business, government, the professions, and certain academic disciplines, the process of collaborative writing is largely ignored by faculty outside of writing courses. Clearly, for students to be fully prepared for the workplace, they need to be taught how to collaborate or co-author documents. They also need to know how to use computer networks and telecommunications technology.

In this chapter we share how we designed, implemented, and evaluated this courseware and telecommunications project. We hope that by sharing our process we will help faculty, instructional designers, developers, and trainers design and implement similar collaboration and telecommunications methods in their classrooms or corporate environments.

OVERVIEW OF THE COURSE

While we designed the courseware for use in any college-level course, the first course in which we implemented the courseware was a junior-senior technical writing course entitled Writing in Your Profession. In this course, students generate documents such as resumes and cover letters, memo portfolios, and instructions, as well as conduct a large project. This large project is a feasibility study consisting of a proposal, a progress report, and a final report. In a more traditional classroom, students work individually on the large project and share their rough drafts of these documents in conference groups. Based on our review of research on collaborative writing, we structured the feasibility study assignment so that students would work in groups of three when designing and documenting their feasibility studies [15, 16].

We also knew that students would need appropriate coaching on how to work collaboratively at the different stages of the writing process [17, 18]. Prior to introducing the feasibility study assignment, the instructor led discussions on the pros and cons of writing together. She talked about textual, individual, and social perspectives on writing, the differences between cooperation and collaboration, alternative strategies to follow when collaborating, how to classify and appoint group tasks, and specific problems to expect at the beginning, middle, and end of the project [19, 20]. Throughout the project the instructor helped students discuss and develop a common understanding about seven variables which Lisa Ede and Andrea Lunsford note influence success and satisfaction in collaborative writing groups: individual *control* over texts, how *credit* will be given, how writers will respond to *modifications* to their texts, *procedures* for resolving disputes, the degree of *flexibility* with pre-established formats, the deadlines and other outside *constraints*, and the *status* of the project in the organization, which in this case was the technical writing course [10]. In addition to this in-class coaching, the courseware itself contained strategies on how to collaborate at the planning, drafting, revising, and packaging stages of the writing process.

OVERVIEW OF THE COLLABORATIVE WRITING COURSEWARE

Collaborative Writing (CW) is an interactive learning and productivity tool that teaches students how to collaborate throughout the writing process and write specific nonacademic documents. CW allows students the flexibility to integrate knowledge acquisition from tutorials with a word processing tool for practice and performance.[1]

From the CW main menu students can choose to read about the definition, types, and stages of, as well as hints about collaboration (Figure 1). Also from the CW main menu students can learn how to write memos, instructions, proposals, short reports, and formal reports. Once students choose a particular tutorial section (for example, Writing Formal Reports), the tutorial menu appears, and a note-taking facility automatically opens so that students can interactively take notes from the tutorials (Figure 2). Throughout the tutorial menus we followed the logical structure of document development—planning, drafting, revising, and packaging. For example, students can choose Audience under the Plan menu, read the tutorial information and generate notes (Figure 3). Question prompts throughout the tutorials help students generate ideas for

---

[1] This software runs on the Apple™ Macintosh 512K Enhanced, Macintosh Plus, Macintosh SE, and Macintosh II. The courseware takes full advantage of the Macintosh user interface and utilizes pull-down menu and windowing capabilities. We selected the Apple Macintosh because of its intuitive user interface. CW's Word Processor is basic, fully integrated with the tutorial, and allows students to easily export all documents they create to any commercial Macintosh word processing system for final packaging of their documents. CW's basic edit functions consist of cutting, pasting, copying, single- and double-spacing of text, setting margins and tabs, saving, and printing.

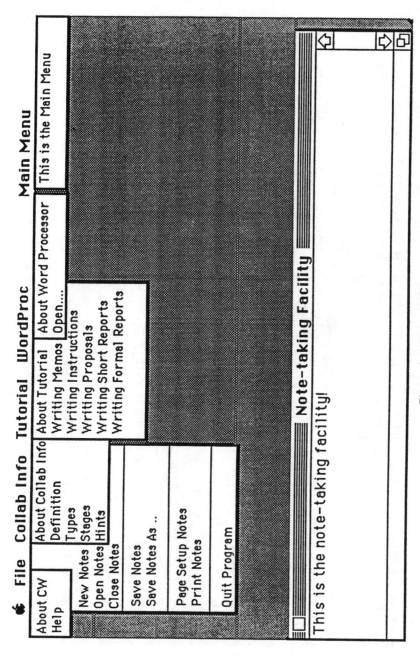

Figure 1. CW main menu.

149

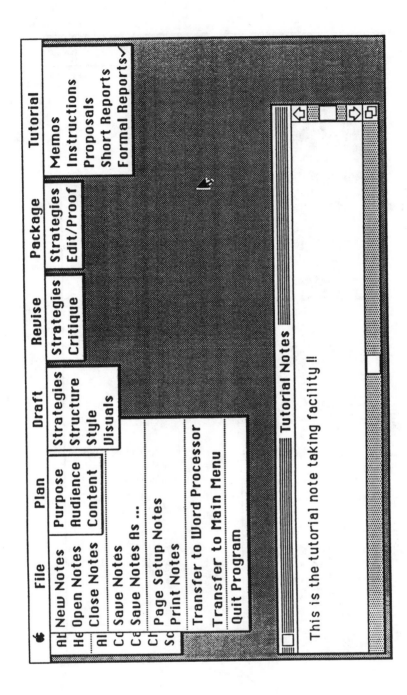

Figure 2. CW tutorial menu and note taking facility.

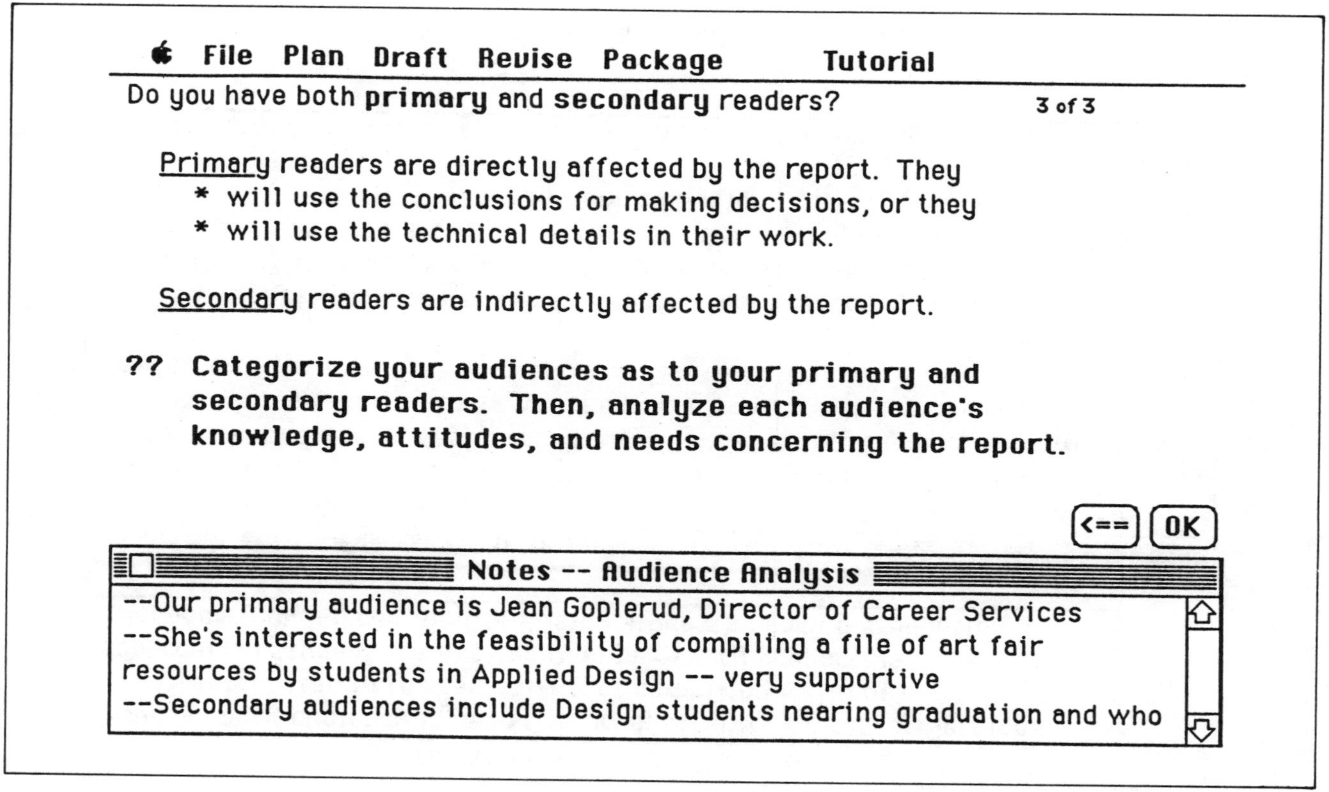

File Plan Draft Revise Package   Tutorial

Do you have both primary and secondary readers?

Primary readers are directly affected by the report. They
* will use the conclusions for making decisions, or they
* will use the technical details in their work.

Secondary readers are indirectly affected by the report.

?? Categorize your audiences as to your primary and
secondary readers. Then, analyze each audience's
knowledge, attitudes, and needs concerning the report.

<== OK

═══ Notes -- Audience Analysis ═══

--Our primary audience is Jean Goplerud, Director of Career Services
--She's interested in the feasibility of compiling a file of art fair
resources by students in Applied Design -- very supportive
--Secondary audiences include Design students nearing graduation and who

Figure 3. Screen and notes from tutorial—formal reports.

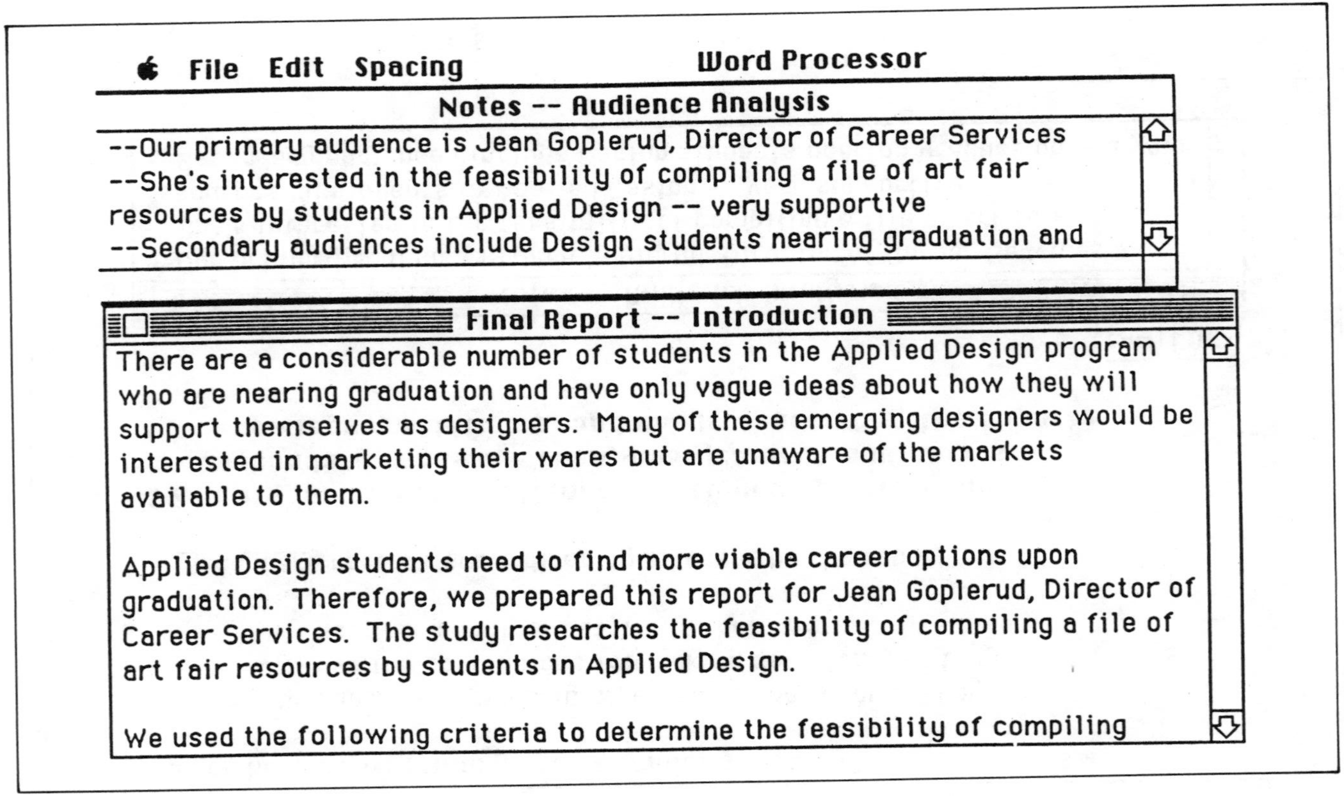

♣ File Edit Spacing   **Word Processor**

Notes -- Audience Analysis

--Our primary audience is Jean Goplerud, Director of Career Services
--She's interested in the feasibility of compiling a file of art fair resources by students in Applied Design -- very supportive
--Secondary audiences include Design students nearing graduation and

Final Report -- Introduction

There are a considerable number of students in the Applied Design program who are nearing graduation and have only vague ideas about how they will support themselves as designers. Many of these emerging designers would be interested in marketing their wares but are unaware of the markets available to them.

Applied Design students need to find more viable career options upon graduation. Therefore, we prepared this report for Jean Goplerud, Director of Career Services. The study researches the feasibility of compiling a file of art fair resources by students in Applied Design.

We used the following criteria to determine the feasibility of compiling

Figure 4. Incorporation of notes into a document.

their first drafts of the actual documents. We designed these writing prompts according to a process orientation toward generating nonacademic documents [17, 21, 22]. Students can at any time take their notes generated in the CW tutorial, rework them in the CW word processor, or incorporate them into existing documents (Figure 4).

In short, students using this courseware analyze both their writing processes and their collaborative strengths and weaknesses, determine how they will approach collaborative projects, and work together to generate information, research a topic, analyze potential audiences, and draft, revise, and package resulting documents.

THE DESIGN PROCESS

Our efforts to create CW followed the same collaborative philosophy that the courseware promotes. We developed CW as part of a collaborative team consisting of a design coordinator, a content expert, a technical communicator, and a programmer.[2] During the first part of the design process the content expert determined what the goals of the courseware package were, and the design coordinator, content expert, and technical communicator developed a flow diagram to serve as the blueprint for the courseware. From this blueprint the programmer was able to build the courseware shell and word processor utility. While programming was occurring, the content expert developed tutorial content to incorporate into the program. A majority of our design team met weekly, and we interacted with an instructional design consultant at major milestones in the program's development. It took us nine months to design and implement the courseware, and throughout this process we made numerous refinements and revisions in program flow, screen design, and tutorial content. We also pilot tested the courseware on students enrolled in the writing course described earlier. This pilot testing allowed us to make further revisions to the courseware, write our documentation, and better learn how to implement the courseware.[3]

CAMPUS-WIDE TELECOMMUNICATIONS

The microcomputer labs at the University of Minnesota are on a university-wide network, and each lab has one machine that acts as a file server or a storage place for computer applications and student and instructor

[2] We developed CW as part of a collaborative team consisting of a design coordinator (Mark DeBower), a content expert (Ann Hill Duin), a technical communicator (Linda Jorn), and a programmer (David Johnson). We wish to thank David Johnson for his programming and testing efforts.

[3] We secured a grant from Apple Computer for hardware for this project. We also wish to thank the College of Agriculture, University of Minnesota, for grant funds from Project Sunrise, a faculty and student development grant, and the University of Minnesota Academic Computing Services and Systems for their support of this project.

File Edit View Special

3562 class

10 items   4,210K in disk   14,961K available

Instructor   Nature   Human 1   AO   Design

Handouts/Assign.   Human 2   Science   Land   Conferencing

Design

Name	Size	Kind	Last Modified
memo to g and k	1K	document	Sat, May 21, 1988
final feas memo to ann	3K	MacWrite document	Thu, May 19, 1988
FINAL DESIGN FEAS.	57K	MacWrite document	Thu, May 19, 1988
Design/memo from AnnD	1K	document	Wed, May 18, 1988
feas. rough 5/14/88	48K	MacWrite document	Tue, May 17, 1988
feas. rough 5/17/88	46K	MacWrite document	Tue, May 17, 1988
Feas Report due date/fr...	3K	MacWrite document	Mon, May 16, 1988
Design/Fdbk/feas1/Ann D.	3K	document	Sun, May 15, 1988
Design memo 5-14-88	1K	document	Sat, May 14, 1988
feas. rough KPH 5/14	38K	MacWrite document	Sat, May 14, 1988
suzi's changes	38K	MacWrite document	Sat, May 14, 1988
letter trans. rough 5-14	4K	MacWrite document	Sat, May 14, 1988
COST TABLE	4K	MacWrite document	Sat, May 14, 1988
feas. rough 5/12/88	37K	MacWrite document	Fri, May 13, 1988
3562.Feas.final.rep.info	33K	MacWrite document	Fri, May 13, 1988
feasibility rough 5/10/88	34K	MacWrite document	Tue, May 10, 1988
recommendations.1	4K	MacWrite document	Tue, May 10, 1988
suzi's changes	38K	MacWrite document	Sat, May 14, 1988
letter trans. rough 5-14	4K	MacWrite document	Sat, May 14, 1988
COST TABLE	4K	MacWrite document	Sat, May 14, 1988
feas. rough 5/12/88	37K	MacWrite document	Fri, May 13, 1988
3562.Feas.final.rep.info	33K	MacWrite document	Fri, May 13, 1988
feasibility rough 5/10/88	34K	MacWrite document	Tue, May 10, 1988
recommendations.1	4K	MacWrite document	Tue, May 10, 1988
progress frnt mat memo	1K	document	Tue, May 3, 1988
Design.prog. second eval	8K	document	Thu, Apr 28, 1988
rough feas. rep. 4-15-88	5K	document	Thu, Apr 28, 1988
memo to A. on prog rep.	1K	document	Tue, Apr 26, 1988
Design / progrep / eval..	9K	document	Tue, Apr 26, 1988
progress rpt draft	7K	document	Mon, Apr 25, 1988
DESIGN.prop.evaluation	7K	document	Tue, Apr 19, 1988
proposal 4/15/88	5K	document	Fri, Apr 15, 1988
Design.prop.comments/Ann	5K	document	Thu, Apr 14, 1988

Figure 5. Design student's access to the file server.

files.[4] Each file server runs AppleShare networking software, and students on this network can send and receive electronic messages among their collaborative team members.

In order to organize the flow of documents between the instructor and students and to ensure the confidentiality of students' messages and documents,

[4] The University of Minnesota microcomputer labs are on an AppleTalk Internet. An Internet is a collection of several local area networks (LANs) which form one large network. The AppleTalk Internet is made up of a number of Apple Macintosh microcomputer laboratories connected over the university's phone system's LANmark Ethernet service.

four types of folders are created on the file server: groups, instructor, handouts, and conferencing folders. Group folders are arranged according to collaborative groups, and one folder is created for each group and assigned the name of the group (Figure 5). Only group members and the instructor have access to these folders and their documents. For the instructor folder, students can send their documents to the instructor, but only the instructor can access documents in this folder. For the handouts folder, students can access all handouts and assignments needed for the class, and the instructor can leave feedback to questions students send concerning the use of computers or the assignments. For the conferencing folder, students leave messages and drafts for students outside of their own collaborative group.

When students want to send a document to team members or to the instructor, they follow a simple log on procedure and an icon representing the file server appears on the screen (in Figure 5 this icon is labeled Hero). Students open this icon and see the four types of folders. When they want to send team members and the instructor a document, they simply drag the document to the appropriate group folder as well as to the instructor folder. If they wish to see if team members have left them any messages or drafts of documents, or if they wish to access an earlier message or document, they open the appropriate group folder.

IMPLEMENTING THE COURSEWARE
AND TELECOMMUNICATIONS

In order to help students learn how to use the technology, and to let these tools enhance the course, the technical writing course meets in our microcomputer lab (Figure 6). As an introduction to the courseware, the instructor models how he or she might use CW to analyze potential audiences, purposes, and content for the various nonacademic documents. The instructor can access specific screens and highlight key questions to answer, lecture from information on specific screens as a type of advance organizer, and direct students to questions pertinent to small group discussions. The instructor makes sure to model how specific responses to question prompts under the Drafting Menu can be reworked into the first draft of the particular document.

As an introduction to the telecommunications, the instructor models how to use the file server to send and receive messages, documents, and feedback. The instructor composes brief messages to students and drags these to the folders, thus showing how he or she will give feedback to students and coach them throughout the collaborative process. To allow the instructor to monitor the students' composing processes, students send all drafts to the instructor folder, and the instructor writes feedback within the document or on separate electronic messages. The instructor then sends the feedback or message to the appropriate group folder. Students access their documents or messages from their group folders and revise and repeat this process.

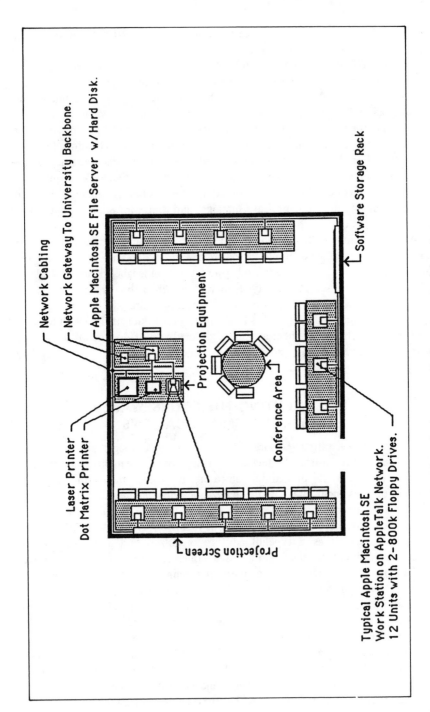

Figure 6.   Rhetoric department microcomputer laboratory.

156

EFFECTS OF THE COURSEWARE
AND TELECOMMUNICATIONS

In order to evaluate this project, we explored the effects of the courseware and telecommunications on students' collaborative processes, the quality of their final documents, and their attitudes toward collaboration (Table 1). Twenty-one college juniors and seniors in the technical writing course were involved in this project, and the duration of the project was seven weeks. On the first day of the course, students were asked to indicate their computer experience. In terms of computer experience, 25 percent of the students described themselves as inexperienced users, 45 percent as intermediate users, 30 percent as experienced users, and none described themselves as advanced users.

Collaboration Logs

Selfe and Wahlstrom, in an article on "casting a broader net with theory and research" on computers and writing, urge researchers to study the sorts of information exchanged when students collaborate on writing efforts that involve computers. They specifically ask researchers to study whether students exchange more information about hardware and software than about the content or organization of the text, and about the effects that collaborative exchanges have on students' composing processes [23]. Therefore, in order to learn about students' collaborative processes and their exchange of information, we asked

Table 1. Research Questions and Measures on the Effects of the
Courseware and Telecommunications on Students' Writing
Processes, Products, and Perceptions

Research Questions	Measures
What strategies do students use when they collaborate?	Collaboration logs
What processes do students follow when they collaborate?	Transcripts of one group's in-class collaborative sessions
In what ways are computers used during the collaborative process?	Collaboration logs
	Volume reports from the file server
	Messages on the file server
How does collaboration affect the quality of subsequent documents?	Ratings from independent professional and academic raters
What are students' attitudes toward the courseware and telecommunications?	Attitude survey

students to record all collaborative activity in collaboration logs (Appendix A). The instructor collected these logs weekly and responded to students' concerns via telecommunications.

These logs gave us a greater understanding about the vocabulary students used to describe their collaborative strategies and where they needed coaching on how to revise their strategies [10]. In these logs, students used collaboration and writing process terms that were given in the courseware tutorials, and they often discussed their insights and frustrations in terms of the specific variables that influence collaboration. The logs helped the instructor to monitor the main tasks and goals students accomplished during their collaborative sessions, to help students assume various roles throughout the process, to assist with frustrations related to the technology, and to coach students through collaboration and writing problems.

Students indicated that in addition to their collaboration via telecommunications, the groups met formally outside-of-class an average of 12.5 times during the seven weeks. In nearly every case the entire group met together. In terms of the stages of the writing process, the groups spent 30 percent of their time planning, 30 percent on drafting, 30 percent on revising, and 10 percent on packaging documents. The groups used the computer for 90 percent of their writing tasks, and of the computer technologies, they used the word processor and telecommunications for roughly 90 percent of their work.

From these logs, we found that students appreciated the use of technology over traditional texts, that the courseware had brought the instructor and students into a common writing workspace, and that the sharing of writing over the network helped students to develop new strategies for collaborating and sharing information about their writing processes and written products. These results parallel those of Selfe and her colleagues who found that the use of microcomputers brought students and faculty into "communal writing spaces," encouraged them to establish new patterns of sharing information about writing, and altered the social patterns that had previously controlled their exchange of written copy [24, 25].

Transcripts

Researchers have investigated group writing strategies by studying the language of writing groups. They have found that a high proportion of students' statements focus on the content of writing and on directives about writing, and that these directives most often follow the assignment constraints set by the instructor. These researchers suggest that writing groups should be encouraged to move away from instructor-initiated talk and instead initiate spontaneous peer talk during their group meetings [26-28].

To learn more about students' collaborative processes via their face-to-face group meetings, Linda Jorn followed one collaborative group, taping their in-class sessions and transcribing their collaboration. We then did a content analysis

Table 2. Types of Collaborative Statements

Statement Types	Definition
Strategy	Members assign responsibilities for completing a project
Audience	Members discuss information relevant to their audience(s) for the project
Verification	Members question, clarify, elaborate, revise, accept, or second guess another member's statement
Content	Members discuss the content, development, and organization of the project
Technology	Members discuss information they need to know in order to use computer technology to produce, revise, or send their documents
Off-task	Members discuss information that is not relevant to the project

on the transcripts using a scale developed by collapsing scales from previous research on collaboration [4,10,29,30].[5] We analyzed six types of collaborative statements: strategies, audience, verification, content, technology, and off-task (Table 2). We analyzed each collaborative member's statement according to this scale and tallied the total number of each type of statement per each collaborative session. We also looked for statements that consistently occurred and that could not be categorized, but might indicate a part of the collaborative process that has not yet been considered by researchers.

These in-class collaborative sessions began one and a half weeks after this group's initial planning for their feasibility study project, and most of the group's discussion concerning their audience for the project occurred outside of class (Table 3). This accounts for the low percentages under Audience. As can also be seen, this group initially spent more time discussing the strategies they would use to collaborate and how to use the technology. However, by the fourth session the group worked on revising their final feasibility report and spent a large amount of time verifying the content and using the technology. The sixth session consisted of the final packaging of the feasibility report. The slight increase in strategy comments during the sixth session reflects the group's emphasis on the final tasks each member did to get the report finalized. This final session also dealt with verifying details, and the majority of these verification statements consisted of discussion on the final content.

---

[5] These researchers describe types of collaborative statements, but they each use slightly different terms for their descriptions. Indeed, Ede and Lunsford note the difficulty inherent in the use of different vocabularies by various researchers and collaborators to describe similar processes.

Table 3. Percentages for Types of Collaborative Statements

Session Number	Strategy	Audience	Verification	Content	Technology	Off-task
1	26%	3%	35%	7%	22%	7%
2	9%	3%	64%	16%	2%	6%
3	16%	2%	70%	4%	2%	6%
4	1%	0%	59%	11%	24%	5%
5	3%	0%	82%	10%	4%	1%
6	7%	0%	64%	12%	10%	7%

As a whole, this analysis shows a large amount of verification. This group spent an average of 62 percent of their time verifying information, or questioning, clarifying, elaborating, revising, accepting, or second guessing each other's statements. This group felt that in order to collaborate effectively, they needed to continually clarify and restate their individual views concerning the organization and wording of the drafts. This amount of verification added to the time needed to complete the documents; however, from their collaboration logs, we know that the group felt that such verification added to the overall quality of their work.

These results parallel Gere and Abbott's results in that the students spent the large majority of their time verifying content and offering directives about writing [27]. The low percentage of in-class time spent on discussing strategies, audience, specific content suggestions, and computer technology may have been due to the use of telecommunications. These students had used the courseware and telecommunications outside of class time to strategize about their audience for the project and to compose their individual parts of the collaborative document. Thus, these in-class, face-to-face meetings became a place to spend the majority of time verifying messages and documents they had sent throughout the week.

Memos on the Course and Collaboration

During the fourth week of the project, the instructor asked the students to send feedback (electronic messages) in which they responded to the course content, collaboration, conferencing, and the use of computers. These memos were not graded and did not fulfill a specific course requirement. Rather, they were intended to elicit responses in which students integrated their insights and frustrations about the course and the use of technology. The following are specific responses concerning the courseware and telecommunications:

I can honestly say that this is the first class I have taken in which I enjoy the group-work experience. Telecommunications makes collaboration simple and fun. The ability to leave memos to other group members makes collaboration possible even when our group can't schedule time to meet.

I think that it is good that everything has to be done on the computer even though it gets frustrating sometimes. It forces us to keep at it, to work out our mistakes and not give up. We are all learning tremendously as far as discovering what a computer is capable of helping us do.

I haven't experienced such a complete use of computers in any class before—even in my computer classes!

All of the students responded positively toward the four areas and gave useful feedback which we implemented when updating the courseware and telecommunications. While we expected some negative feedback, we did not receive any, and this possibly was due to the collaboration logs in which students had already expressed some specific frustrations with the technology.

Messages on the File Server

A number of studies to date on electronic messages have been conducted in industry settings. These researchers have found that the majority of messages have centered on setting up meetings or other logistical needs [31, 32]. Kiesler and her colleagues have explored how people participate in electronic conferences and how this affects group efforts to reach consensus. They found that group members using electronic messages participated more equally than they did when they talked face to face, were more uninhibited than they were in face-to-face groups, and had less chance that one person in the group would dominate the discussion. They also found that in situations where electronic messages were the only form of communication, that the messages were less personal and were directed away from one's audience [33, 34].

For this project, in order to monitor and coach students through the writing process, on a daily basis the instructor checked the groups' electronic messages to each other and for herself. At the beginning of the course the instructor made overheads of some of the messages and shared these in class. This activity inspired other groups of students to begin taking advantage of telecommunications as a means to give and receive feedback and other messages. Later in the course the instructor only displayed students' messages when they pinpointed concepts that were crucial to designing and documenting the feasibility study documents. The following are examples of messages on the file server:

Student-to-Student Message

TO: Darcy and Chris
FROM: Deb
SUBJECT: Response to Student Survey

I think our survey looks good, but I'm wondering if it would be more clear if we went into more detail about what a financial counseling service would offer. The following are changes I think we should make: (a list of changes follows.) What do you think?

Student-to-Instructor Message

TO: Ann
FROM: Janae and Kevin
SUBJECT: Questions Dealing with our Final Report

We are having some problems separating the information that should go in a factual summary from that which belongs in annexes. We're not sure where to draw the line as far as detail goes.

Please access our Final Feas Report file and give us some feedback. We think everything else is under control.

Instructor-to-Student Message

TO: Agriculture Group
FROM: Ann
SUBJECT: Feedback on Draft 2 of Your Feasibility Report

Nice revision of your first draft! The following is my feedback on your second draft. In addition to this list, please meet with me for about 30 minutes and I'll help you read through your draft with an eye for editing concerns. Let me know if you can meet with me next Tuesday afternoon. (List of feedback follows.)

Students sent inquiries, responses, information, and directives to each other and to their instructor. As can be seen from these messages, the majority of the content centered on the content and form of their collaborative documents.

Volume Reports

We were also interested in how much the students used the telecommunications network. Therefore, we downloaded volume reports from the file server, or information concerning the number of times groups accessed and used the file server on a weekly basis. Specifically, each week we obtained the following data from the server: total kilobytes of use, total files on the server, the average number of files per group, the number of files in the instructor folder, and the size of the files. This data helped us chart trends of the groups' use of the server.

We found that the students' use of the server increased dramatically over the seven week period. By the end of the project, the twenty-one students and instructor had written, sent, and/or received 2767 kilobytes of information. Since one page of text is approximately 3 kilobytes, this translates to approximately 922 pages of text or forty-four pages per student. By the end of the project, each group had an average of forty-two files on the server, or about fourteen files per student, and the instructor had written, sent, and/or received 277 files (drafts, messages, and feedback). The size of the files also increased throughout the project as students progressed from writing shorter proposals and progress reports to writing longer final reports. This increase in file size also supports the fact that students used the network for more than brief logistical messages to each other; they indeed composed, shared, and evaluated their own and their peers' writing via telecommunications.

Ratings from Independent Raters

In order to evaluate documents from the collaborative groups, printouts of two components of the feasibility study—final drafts of proposals and final drafts of the final reports—were collected from all seven groups. These drafts were randomly ordered within the document type, and two pairs of raters—one pair of composition instructors and one pair of professionals from industry—rated the documents. The raters evaluated the documents according to a four point holistic scale modeled after the suggestions of White with 1 being a poor quality document and 4 being a document of excellent quality [35]. The composition instructors gave the proposals a mean rating of 2.56 and gave the feasibility reports a mean rating of 3.29. The professionals gave the proposals a slightly higher mean rating of 2.86 and gave the feasibility reports a mean rating of 3.14. These ratings showed basic agreement in the instructors' and professionals' ratings as well as students' improvement in writing higher quality documents as the project progressed.

One naturally expects students' writing to improve across the duration of a course. One would especially expect students' writing to improve if they repeated a similar type of writing and writing process; that is, if they wrote several proposals in succession. However, in this course, each successive document was new to the students. Thus, the final feasibility report was itself a new type of document with added style and content constraints. With this in mind, we see the higher ratings on the final reports as significant. The final reports involved collaborating and coordinating efforts to design and document five weeks of investigative material for a real client, and independent raters saw these final documents as quality documents.

We were also happy to see that the use of computers and telecommunications as a means to collaborate did not adversely affect the quality of students' writing. Indeed, the raters, in written comments below their holistic ratings, indicated that the overall quality of the writing was good and that each document displayed a uniform style rather than three separate styles put together.

Attitude Survey

Attitude surveys have been widely used to study collaborative strategies and writing strategies in the workplace [10, 36]. Thus, we developed an attitude survey to get information concerning students' views toward the course, collaboration, and the use of the courseware and telecommunications. This survey contained both positively and negatively phrased statements and employed Likert scales. We also included open-ended questions to gain more specific information on the use of the technology. In designing this survey, we used statements that Ede and Lunsford reported in their research of professional writers [10]. Specifically, we asked students to indicate their attitudes toward the seven variables that had been introduced to them via the courseware—control, credit, modifications, procedures, flexibility, constraints, and status—and we asked students to indicate the collaborative strategies they used most often.

Overall, the students responded positively toward collaboration and computers. Nearly all of the students (90%) enjoyed working in groups on the project, 85 percent were satisfied with their collaborative experiences, and 90 percent were satisfied with the documents they produced. The students' responses to the seven variables that affect members' attitudes toward collaboration were also positive and this correlated with their positive responses toward collaboration as a whole. The procedures variable was the only one with which students had difficulty; 70 percent of the students were undecided or disagreed with the statement concerning whether their group had established procedures for resolving disputes. Students felt that the instructor needs to help groups deal with procedures for resolving conflict and verifying information. Students indicated that they used numerous strategies to collaborate, but the most frequently used strategy by far (80%) was to have the group plan together, to have each member draft specific parts of the document, and then to compile the parts and revise as a group.

The many uses of computers in the class did not produce negative attitudes and this validates our previous research on students' attitudes toward enhanced uses of computers [37]. A strong majority of the students (85%) felt that computers gave them more time to revise, and 100 percent indicated that telecommunications was especially helpful in sharing and receiving feedback from the instructor.

On the open-ended questions, students wrote that in terms of collaborating, they appreciated learning from others, liked sharing the workload, and enjoyed sharing information via the computer. As for what was hardest about collaborating, students wrote about the difficulty in giving good feedback, working together during stressful times, agreeing on how to word a document, blending the material into a uniform style, getting equal participation on the project, and delegating tasks. Students wrote that the best part about using computers was the ease of manipulating text and the increased time for revising and improving texts. As for what they liked least about using computers, students wrote about the lack of access time in the campus labs and the chance of losing or misplacing files.

CHANGING ROLES—
SOCIALIZING THE WRITING PROCESS

Enhanced uses of existing technology have the capacity to socialize the writing process and to change teaching and learning. Through appropriate courseware, we can capture the computer's capacity to present the writing and collaborative processes in one medium. Through appropriate courseware design, we can free users from linear applications and allow users to direct their own learning. Through telecommunications, the instructor no longer waits for students to come and ask questions; students access that instructor's knowledge anytime. Writing, learning, and collaboration can become social processes emulating from continuous dialogues.

Instructor as Coach

In this project we found that when an instructor uses courseware and telecommunications to enhance a course, that instructor lectures less and coaches more than in a traditional classroom. While we did not do a formal analysis of the instructor's feedback and messages to the groups, the sheer amount of telecommunications interaction promoted students' increased writing, revision, and in-depth analyses of their audiences, topics, and collaborative strategies. Arms, when analyzing how instructors respond to using computers, notes that "the surprise is that teachers are moving from the role of pedagogue to that of advisor, from teaching writing to fostering creativity. . . . The computer can put the teacher in the role of a Socratic prodder rather than an error finder" [38, p. 74]. Indeed, in this project the instructor's roles were many: designer, coach, tutor, advisor, diagnoser, discoverer, reactor, and even a collaborator with the groups.

Technology as Coach

The courseware itself and the integration of the courseware in the computer classroom also contributed to teaching and learning. At the design stage, we found the courseware's content and organization to evolve from the interaction between the content expert's experience in coaching collaborative writing groups and the other team members' expertise in designing intuitive and usable courseware. At the implementation stage, students did not view the courseware as an outside-of-class assignment but rather as an aid to help them discuss and enhance their writing and collaborative processes.

The use of telecommunications also contributed to the teaching and learning. The instructor did not have to wait for the next class period to deliver pertinent information to students, and students did not have to wait until the instructor's office hours or the next class period to ask important questions. Students could send and receive feedback or collaborate and cooperate with team members and the instructor at any time or stage of their writing processes. We found that the courseware and telecommunications helped students confront the complexities of collaborating described by Kraut et al.: developing an equitable division of labor, subtly supervising peers, sharing ill-formed information, and coordinating writing that is continually evolving [39].

Thus, the technology enhanced teaching and learning as the students and the instructor viewed writing as a social process. Computers facilitated collaboration as students shared not only their final written products but also their processes and their struggles—their documents in the making. Computers facilitated conferencing as the instructor, instead of looking only at the written draft of a document and finding it difficult to diagnose, advise, and instruct the writers, came much closer to the students' actual writing events [40].

Socializing the Classroom

Lyman, in a recent article on the computer revolution in the classroom, asks the following questions:

- What's happening to the relationship between faculty and students in the computerized classroom?
- What is changing in the computerized classroom?
- What is it about computing that has the capacity to change teaching and learning? [2, p. 26]

We believe that courseware and telecommunications have the capacity to change faculty and student roles in the learning process. When courseware concentrates on group development of nonacademic documents, it becomes a catalyst to prepare our students for writing in the workplace. When collaboration is implemented via telecommunications, students actively give and receive feedback, listen and verify information.

What is changing in the computerized classroom? Instructors and students are becoming engaged in active learning roles. Computers are redefining work, refining writing, and redesigning communication. Students are seeing writing as a social, problem-solving activity, and they will take this knowledge with them into the workplace.

APPENDIX A: COLLABORATION LOG

Date _____   Name _____   Number of collaborators _____

What stage of the Collaborative Writing Process are you in?

_____ Planning   _____ Drafting   _____ Revising   _____ Packaging

What was your main task or goal during this collaboration?

What strategies did you use to meet your goals?

What was your main role? (e.g., motivator, listener, idea generator, note taker, task delegator)

Insights or frustrations?

What technologies did you use during this collaboration?

____ Phone ____ Pen and Paper ____ Chalkboard ____ Computer ____ Other

If you used the computer, what did you use the computer for?

____ Word Processing   ____ Tutorial   ____ Graphics
____ Telecommunications   ____ Other

REFERENCES

1. K. A. Bruffee, Writing and Reading as Collaborative or Social Acts, in *A Sourcebook for Basic Writing Teachers*, T. Enos (ed.), Random House, New York, pp. 565-574, 1987.
2. P. Lyman, The Computer Revolution in the Classroom: A Progress Report, *Academic Computing, 2*:6, pp. 18-20, 43-46, 1988.
3. C. Daiute, Issues in Using Computers to Socialize the Writing Process, *Educational Communication and Technology Journal, 33*:1, pp. 41-50, 1985.
4. N. Allen, D. Atkinson, M. Morgan, and T. Moore, What Experienced Collaborators Say about Collaborative Writing, *Journal of Business and Technical Communication, 1*:2, pp. 70-90, 1987.
5. P. Anderson, What Survey Research Tells Us about Writing at Work, in *Writing in Nonacademic Settings*, L. Odell and D. Goswami (eds.), The Guilford Press, New York, pp. 3-38, 1985.
6. L. Faigley and T. Miller, What We Learn from Writing on the Job, *College English, 44*:6, pp. 557-569, 1982.
7. J. Paradis, D. Dobrin, and R. Miller, Writing at Exxon ITD: Notes on the Writing Environment of an R & D Organization, in *Writing in Nonacademic Settings*, L. Odell and D. Goswami (eds.), The Guilford Press, New York, pp. 281-307, 1985.
8. L. Odell, Beyond the Text: Relations Between Writing and Social Context, in *Writing in Nonacademic Settings*, L. Odell and D. Goswami (eds.), The Guilford Press, New York, pp. 249-280, 1985.
9. S. Doheny-Farina, Writing in an Emerging Organization: An Ethnographic Study, *Written Communication, 3*:2, pp. 158-184, 1986.
10. L. Ede and A. Lunsford, Why Write . . . Together: A Research Update, *Rhetoric Review, 5*:1, pp. 71-81, 1986.
11. _____, Collaborative Learning: Lessons from the World of Work, *Writing Program Administrator*, in press.
12. S. B. Kiesler and L. S. Sproull, *Computing and Change on Campus*, Cambridge University Press, New York, 1987.
13. I. Greif (ed.), *Computer-Supported Cooperative Work*, Morgan Kaufmann Publishers, Inc., San Mateo, CA, 1988.
14. A. S. Bednar and R. J. Olney, Communication Needs of Recent Graduates, *The Bulletin, 4*, pp. 22-23, December, 1987.
15. D. W. Johnson and R. T. Johnson, Computer-Assisted Cooperative Learning, *Educational Technology, 26*:1, pp. 12-18, 1986.
16. D. Trowbridge, An Investigation of Groups Working at the Computer, in *Applications of Cognitive Psychology: Problem Solving, Education, and Computing*, D. Berger, K. Pexdek, and W. Banks (eds.), Lawrence Erlbaum Associates, New Jersey, pp. 47-58.
17. D. H. Barbour, Process in the Business Writing Classroom: One Teacher's Approach, *The Journal of Business Communication, 24*:1, pp. 61-64, 1987.
18. J. Forman and P. Katsky, The Group Report: A Problem in Small Group or Writing Processes?, *The Journal of Business Communication, 23*:4, pp. 23-35, 1986.

19. L. Faigley, Nonacademic Writing: The Social Perspective, in *Writing in Nonacademic Settings*, L. Odell and D. Goswami (eds.), The Guilford Press, New York, pp. 229–248, 1985.

20. S. M. Hord, A Synthesis of Research on Organization Collaboration, *Educational Leadership, 43*:5, pp. 22–26, 1986.

21. L. Brannon, Toward a Theory of Composition, in *Perspectives on Research and Scholarship in Composition*, B. W. McClelland and T. R. Donovan (eds.), Modern Language Association, New York, pp. 6–25, 1985.

22. J. Trimbur, Collaborative Learning and Teaching Writing, in *Perspectives on Research and Scholarship in Composition*, B. W. McClelland and T. R. Donovan (eds.), Modern Language Association, New York, pp. 87–109, 1985.

23. C. L. Selfe and B. J. Wahlstrom, Computers and Writing: Casting a Broader Net with Theory and Research, *Computers and the Humanities, 22*:1, pp. 57–66, 1988.

24. C. L. Selfe and J. D. Eilola, The Tie That Binds: Building Discourse Communities and Group Cohesion through Computer-Based Conferences, *Collegiate Microcomputer, 6*:4, pp. 339–348, 1988.

25. C. L. Selfe and B. J. Wahlstrom, An Emerging Rhetoric of Collaboration: Computers, Collaboration, and the Composing Process, *Collegiate Microcomputer, 4*:4, pp. 289–294, 1986.

26. A. DiPardo and S. W. Freedman, Peer Response Groups in the Writing Classroom: Theoretic Foundations and New Directions, *Review of Educational Research, 58*:2, pp. 119–149, 1988.

27. A. R. Gere and R. D. Abbott, Talking About Writing: The Language of Writing Groups, *Research in the Teaching of English, 19*:4, pp. 362–385, 1985.

28. C. P. Walker and D. Elias, Writing Conference Talk: Factors Associated with High- and Low-Rated Writing Conferences, *Research in the Teaching of English, 19*:4, pp. 266–285, 1987.

29. R. Y. Hirokawa, Why Informed Groups Make Faulty Decisions—An Investigation of Possible Interaction-Based Explanations, *Small Group Behavior, 18*:1, pp. 3–29, 1987.

30. L. A. Suchman and R. H. Trigg, A Framework for Studying Research Collaboration, *Proceedings of the Conference on Computer Supported Cooperative Work*, Austin, Texas, pp. 221–228, 1986.

31. J. Sherblom, Direction, Function, and Signature in Electronic Mail, *The Journal of Business Communication, 25*:4, pp. 39–54, 1988.

32. R. B. Mitchell, M. C. Crawford, and R. B. Madden, An Investigation of the Impact of Electronic Communication Systems on Organizational Communication Patterns, *The Journal of Business Communication, 22*:4, pp. 9–16, 1986.

33. S. Kiesler, J. Siegel, and T. W. McGuire, Social Psychological Aspects of Computer-Mediated Communication, *American Psychologist, 39*:10, pp. 1123–1134, 1984.

34. L. Sproull and S. Kiesler, Reducing Social Context Cues: Electronic Mail in Organizational Communication, *Management Science, 32*:11, pp. 1492–1512, 1986.

35. E. M. White, *Teaching and Assessing Writing*, Jossey-Bass Publishers, San Francisco, 1985.
36. P. V. Anderson, What Survey Research Tells Us About Writing at Work, in *Writing in Nonacademic Settings*, L. Odell and D. Goswami (eds.), The Guilford Press, New York, pp. 3–84, 1985.
37. A. H. Duin and L. A. Jorn, CAI for Pre-Professional Courses: Effects on Students' Processes and Attitudes, Paper Presentation at the College Composition and Communication Conference, St. Louis, March, 1988.
38. V. M. Arms, Engineers Becoming Writers: Computers and Creativity in Technical Writing Classes, in *Writing at Century's End: Essays on Computer-Assisted Composition*, L. Gerrard (ed.), Random House, New York, pp. 64–78, 1987.
39. R. E. Kraut, J. Galegher, and E. Carmen, Relationships and Tasks in Scientific Research Collaboration, *Human-Computer Interaction, 3*, pp. 31–58, 1987–1988.
40. D. Payne, Computer-Extended Audiences for Student Writers: Some Theoretical and Practical Implications, in *Writing at Century's End: Essays on Computer-Assisted Composition*, L. Gerrard (ed.), Random House, New York, pp. 21–26, 1987.

CHAPTER 9

Peer Collaboration and the Computer-Assisted Classroom: Bridging the Gap between Academia and the Workplace

WILLIAM VAN PELT
ALICE GILLAM

If you ask technical writers whether or not they do much collaborative writing, you are likely to be met with a puzzled look. Yet if you question them further about who shares responsibility for the content and style of their documents, you will discover that most of the writing they do is collaborative in one way or another. Similarly, if you ask technical writers about the importance of the computer to their day-to-day writing and collaborations, you will find that its use is so indispensable that they cannot imagine their work without one. Mary Mullins, a technical writer at First Wisconsin National Bank, said of the computer: "It's like an extension of my hands" [1]. None of these responses comes as any surprise. Recent studies of workplace writing note the frequency of collaboration [2-5] and the ubiquity of word processing and tele-communications [6,7].

To help students develop skills needed in the workplace, many technical writing courses now include collaborative assignments in the context of computer-assisted classrooms [8-11]. The University of Wisconsin-Milwaukee technical writing program has included such assignments in both introductory and advanced courses since 1983. Designed to simulate workplace collaborations, these assignments require student writers to accept and offer constructive criticism, to work through multiple review and revision cycles, to gather information from others, to use the computer for communication and coordination of projects, and to produce writing which is a corporate, not an individual, product. Though we have been pleased with the results of these assignments, we have been curious about the degree to which our assignments simulate workplace collaborations and the degree to which they teach identifiable collaborative skills. Nancy Allen and her coauthors note that: "the information we have about collaborative writing" in the workplace "is fragmentary and unfocused" [12, p. 70]. We agree that more detailed knowledge about workplace collaborations is needed; further, we feel that more detailed knowledge about classroom collaborations is equally important [13]. We are interested in how collaborative skills develop as well as in how collaborative writing gets done in the "real world."

Our curiosity led us to undertake two research projects, one designed to glean more information about workplace collaborations, the other to develop new information about classroom collaborations. Because we chose to study the collaborations of students taking an advanced technical writing course which focused on documenting computer software, we surveyed and interviewed professional technical writers who specialize in software documentation. Four questions guided our study of collaborations in both settings:

- What does collaboration entail?
- How does the computer facilitate collaboration?
- What developmental writing and social issues emerge in the process of collaboration?
- What are the theoretical implications of our findings?

The following chapter briefly presents results from our study of the workplace and then concentrates on the results from our study of the classroom collaborations. We conclude by proposing a theoretical model for understanding the relationship between collaborative writing in these two settings.

WHAT DOES WORKPLACE COLLABORATION ENTAIL?

To learn more about the forms, definitions, and practices of collaboration in the workplace, we evaluated the results of a survey of seventy technical writers employed in businesses and industries that produce various forms of computer documentation. The survey was conducted by Charlotte Ruenzel and Judy

Sheldon, two technical writers at Deluxe Data Corporation in Milwaukee who gave us access to their finding [14]. For more in-depth information, we conducted two-hour audiotaped interviews with three writers at large corporations, two of whom are managers of technical writing departments [1, 13, 14]. All subjects identify "technical writing" as their profession and all write primarily software documentation, including memoranda, needs analyses, proposals, user manuals, reference manuals, and tutorial documents for computer programs.

The survey of professional writers confirmed our expectations about the presence of the computer and peer exchange in the workplace: 80 percent reported daily use of the word processor, 76 percent reported research and interviewing as primary activities that usually require collaboration with other members of a writing team, and 70 percent reported "peer editing" as a primary activity which relied on written and oral commentary from fellow writers [14, p. 24]. This survey clearly identifies collaborative writing activities and computer use as important aspects of the technical writer's work routine.

From our interviews of individual writers, we learned that although they rarely use the term "collaborative writing" in the workplace, they agree that the idea of a collaborative effort which requires substantial input from others permeates almost every aspect of the writing process on the job [1, 15, 16]. Like Ede and Lunsford, we discovered that both the concept of collaborative writing and the terminology used to describe it are problematic and that many of our questions about collaboration emerged from a "growing recognition of the dichotomy between current models of the composing process and methods of teaching writing, almost all of which assume single authorship, and the actual situations students will face upon graduation, many of which may well require coauthorship or group authorship" [5, pp. 71-72].

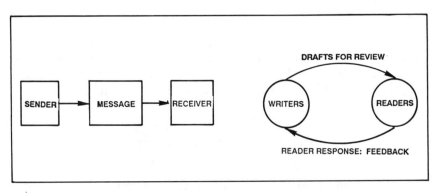

Figure 1. Sender-message-receiver model and
collaborative model of communication.

Our research confirms that the basic model of writing in the workplace is collaborative rather than "single author" and we represent this distinction, in part, through a revision of the traditional sender-message-receiver model of technical communication (see Figure 1). The sender-message-receiver model emphasizes the content of the message itself as a one-way linear communication in which the "sender" originates the message and the "receiver" plays an essentially passive role (see left side of Figure 1). Our revision of this model shows readers as active participants in a cycle of revision and rewriting which incorporates audience awareness and reader response, or "feedback," as critical aspects in the social construction of meaning (see right side of Figure 1). In this model, the collaborative nature of technical discourse emerges from an on-going dialogue between readers and writers that could be called "dialogic" in a Bakhtinian sense which understands linguistic constructs as utterances already populated by the intentions of other speakers.

In reality, collaborative writing appears much more complex than the simple diagrams above suggest. The three professional writers we interviewed identified several forms of collaborative writing, and although the workplace terminology varied from company to company, nearly all the terms fell into two basic functional categories: "writing team work" (working with a "writing team," a "project team," or a "review team"), and "cooperative writing" (writers working cooperatively together as coauthors or co-designers of a project). These two categories closely resemble the distinction Nancy Allen et al. and Harvey Wiener make between collaborative "group work" and "shared-document collaboration" [12, pp. 72-73; 17, pp. 55-57]. In the remainder of this chapter, we use "team work" to encompass "group work" in which authors use input or feedback from others but retain final responsibility for their own writing decisions. We use "shared-document collaboration" for collaborations in which writers share authority and decision-making responsibility for important aspects of the writing process such as planning, composing, and formatting decisions. The fundamental distinction between these two general types of collaboration is that team work leaves the responsibility for the writing with a single author whereas shared-document collaboration requires joint decision making at crucial junctures in the writing process.

Team-Work Collaboration

Team-work collaborations refer to projects in which an individual author maintains authority over final writing decisions but receives varying degrees of input from others (see Figure 2). In this kind of writing, a single author remains at the center of the decision-making process while collaborating with a larger team of contributors and readers. In practice, however, many variations of team work exist. Some teams subdivide a writing project into discrete units which require individual authorship from several members of the team and only

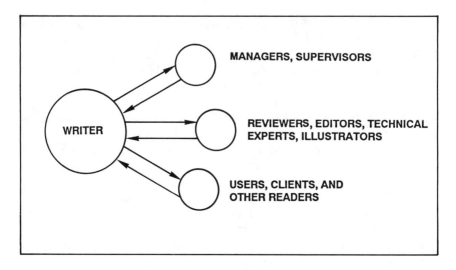

Figure 2. Team-work collaboration.

minimal exchange during the actual writing of each segment while the team leader retains responsibility for final editing and production. Other teams use peer response groups in which writers share ideas and texts and receive substantial feedback from others, but always make final decisions about style and content on their own.

According to the three writers we interviewed, the most common forms of team work are listed below, roughly in descending order from most common to least common:

- submitting documents to supervisors or managers for review, comment, and approval;
- interviewing and interacting with technical staff to establish the scope and generate content for initial drafts and to revise final drafts for accuracy and completeness;
- peer editing and review of writing for style, content, and rhetorical effectiveness; and
- managing a review team or project team to coordinate technical and stylistic input for multiple revisions, final approval, and production of a long document or set of documents.

In each of these cases, a writer may accept critical input from other writers, editors, illustrators, content experts (engineers and technicians), marketing personnel, product specialists, users, and managers, but retain individual responsibility for the final document.

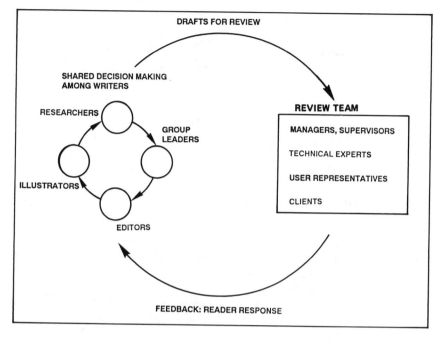

Figure 3. Shared-document collaboration achieved through
shared decision making.

Shared-Document Collaboration

The second and less common category of collaborative writing identified by
the writers we interviewed involves more than one person in the primary
decision-making tasks of planning, writing, and finalizing the document (see
Figure 3). In this model of collaboration, the decision-making authority of the
single writer is replaced by shared authority of several writers, who may or may
not accept input from a larger team of reviewers. Each of the shared-document
collaborators may perform as a contributing writer, a specialist in some aspect of
the writing project (graphics, editing, system design), or both. The crucial
distinction between shared-document collaboration and team-work collaboration
is that writing decisions are shared rather than individual.

Two of our subjects called shared-document collaboration "cooperative"
writing, and Allen et al. specifically defines this kind of collaboration as writing
which "involves collaborators producing a shared document, engaging in substantive interaction about that document and sharing decision-making power and
responsibility for it" [12, p. 70]. As both Wiener's and Bruffee's definitions of
"collaborative learning" imply, the key to this kind of collaboration is
"consensus," or a process of "intellectual negotiation" which leads to "joint

decisions," "collective judgments in groups," and shared authority over the writing process and product [17, pp. 54-55; 18, p. 107]. Among the kinds of workplace collaborations described by our three subjects, the following three examples most closely resembled this kind of shared-document collaboration:

- working cooperatively as a group or committee to establish stylistic and document design standards for the group as a whole or for other groups of writers within a department or company;
- cooperative writing in which two or more individuals plan and design a document or a set of documents together, write individual sections separately, and then work closely together to produce the final document or set of documents in a single format; and
- coauthoring in which writers actually draft or revise the document together.

Workplace writers may not intentionally engage in shared-document writing as a collaborative learning experience, but the struggle for consensus often requires new levels of intellectual negotiation and diplomacy for many writers. As one of the writers who is also a manager of her writing group said: "The goal of this kind of coauthoring or collaborative writing is to produce a manual that looks and reads as though one person wrote it, even though two or three people wrote it together—and this can be difficult to achieve, especially if writers don't learn to communicate well with each other" [16].

Although we clearly distinguish here between "team work" and "shared-document collaboration," these two categories are rarely separate or mutually exclusive in the workplace where the various forms of collaboration frequently overlap. In fact, successful shared-document decision making frequently engages and depends on a good deal of preliminary and continuous team work. For example, when writers coauthor a shared document, they will probably also use most of the techniques listed above under the team work category to gather initial materials from content experts and maintain a collaborative relationship with review teams, technical personnel, and peer editors. Similarly, when a standards committee negotiates to reach a consensus about a company style guide, members must solicit substantial input from writers outside the committee. After reaching consensus based on this research, the committee often negotiates further with editors and managers to finalize company standards.

As Ede and Lunsford suggest, distinctions between single author and multiple author documents become problematic when "professionals regularly make use of in-house or 'boiler-plate' materials, and they may use such materials verbatim ... without acknowledging or documenting their use of this 'silent' coauthor's work" [5, p. 73]. A company's organizational procedures, then, may promote a kind of team work which reduces the need for immediate shared-document decision making by developing pre-established standards for document planning, stylistic conventions, and formatting decisions. Thus a writer unconsciously collaborates with other writers, builds upon their work, and participates in a larger consensual process without overtly engaging in a shared-document decision making. For example, when writers revise existing manuals written by previous employees (who often have left the company or moved on to other positions), the new writers engage in a "silent" coauthoring relationship with the first writer which requires adopting many of the original writer's assumptions and standards or perhaps contending with the original author's text. On rarer occasions, however, new writers may find it necessary to collaborate or contend with the original authors if they are still at the company and available for consultation.

HOW DOES THE COMPUTER FACILITATE WORKPLACE COLLABORATION?

Although the computer was taken for granted as the primary tool for writing which practically no writer would do without, its role in the collaboration process emerged specifically as a medium for facilitating communication. The main uses included: using electronic mail on a daily basis to communicate with other members of the writing team; sharing texts online with other writers, editors, and reviewers; sharing standard document format templates, page layouts, and style sheets online; and retrieving information, such as screen displays for use as visual aids, directly out of technical databases for incorporation into documents. Just as writers use the computer to make writing and revising faster and easier, writers use the computer as a collaborative tool because it increases their productivity by making the communication process faster and easier.

Current research supports our findings. Will Wheeler emphasizes that the "group process" of technical writing benefits greatly from using word processors "to generate the outline and style guide and support the revisions for cohesiveness ... [and] to set document layout characteristics by defining the word processor's 'style sheet.' The writer just identifies the type of text they are producing ... and ... does not have to be concerned with layout at all" [6, p. 39]. Janis Forman emphasizes that electronic mail is "the preferred communication medium when managers need to revise drafts ... and ... it is used throughout the writing process" [7, p. 26]. Frederick O'Hara states that "With electronic mail, it is not uncommon for me to go through three review sessions with a project coordinator in a single working day ... and the result for me is less time wasted producing text that will ultimately have to be replaced" [19, pp. 81–82]. We also found that the computer improves the quality of collaboration primarily by making communication between writers easier, more accurate, and more frequent. Once writers start using the computer in this way, they rapidly become heavily dependent on it, often unconscious of how thoroughly it is woven into their writing process.

WHAT DEVELOPMENTAL WRITING
AND SOCIAL ISSUES EMERGE
IN WORKPLACE COLLABORATION?

Our interviews confirmed that all writing in the workplace is "collaborative" because more than one person is always involved, either as reviewer, content expert, editor, supervisor, coauthor, or some combination of these. But what we found most interesting is the set of social and political issues that shape this collaboration according to the ethos of the corporate world. Specifically, we found that in the workplace, authorship is conceived of as corporate, and group work is driven by productivity.

Collaborative Responsibility:
The Individual Writer and the
Problem of Ownership/Authorship

Everyone knows that when writers complete a users' manual or reference manual for a company or large institution, they rarely, if ever, put their names on the final document as the sole authors. We were therefore intrigued when one of the managers said that her supervisor suggested that writers should put their names on their final documents, but that the technical writers felt this was inappropriate and preferred not to do so. She explained that the writers felt that their writing was a product of corporate effort and that responsibility for it belonged to everyone who had contributed, not just to the individual writer. Besides, if the technical information was incorrect, the content expert who reviewed the document and signed it off would be responsible, not the writer [16].

Ownership is clearly a significant issue: although writers need to take pride in their professional writing ability, they must also learn to practice "ego-less writing" which avoids excessive ego investment in the writing process and submits readily to criticism and suggestions from others. Technical writers must be mature enough to realize that the responsibility for the document rests with many people, that it must satisfy the audiences' needs (not the author's ego) and a specific business or technical function, and that it will undergo constant revision and change. Moreover, not just the authorship, but the liability and professional impact of the final document are also perceived as collaborative responsibilities within the corporation. At the same time, writers must know when to assert their own professional expertise as word-smiths or stylists in order to maintain the readability and rhetorical effectiveness of the document. In addition to good writing skills, then, successful collaboration depends primarily on good communication skills, professional maturity as a writer, and an awareness of the social and political realities of the corporate world.

The Collaborative Process and the Management of Productivity

More than any other factor in the workplace, productivity determines the type of collaboration which managers and writers choose for a project and the amount of time they can devote to the actual collaborative effort. For example, writers must be efficient when dealing with technical personnel since content experts or product support personnel have their own busy schedules and often only grudgingly spend time providing information for a new document or reviewing drafts for accuracy. Writers must therefore become efficient managers of the research, writing, and review process by budgeting time for interviews and parceling out sections of a document for review. The writers and managers we interviewed emphasized that they become better managers of their writing process when they recognize and use estimating, scheduling, and diplomacy as productivity tools. All of these skills affect how much and in what ways writers collaborate with others.

All the writers interviewed indicated that coauthoring is the least practiced form of collaboration because it is the least efficient in terms of total hours spent on a project: it takes more people-hours to produce a document if two people collaborate as primary authors than if one person writes the document with input from others. Only when the writing project is too large for one person to manage alone is coauthoring perceived as a preferred practice. For example, one manager asked two writers to coauthor a large manual because the client wanted the project within eight months, and she knew that one writer working alone would take a year to complete the project. She estimated that two writers could finish the project in seven months and thus meet the deadline. However, she also recognized that coauthoring expended more company resources because the two coauthors would use a total of fourteen people-months rather than the more efficient twelve people-months used by a single author. At the time of our interview, this manager had begun to worry that the coauthoring process might take longer than expected because the two writers were having difficulty communicating and sharing writing decisions. At another company, a coauthoring team of four writers were assigned to write a complete library of documentation for a large computer system. The task required writing ten new manuals and revising fifty-three existing manuals. The coauthoring method was chosen for two reasons: first, each writer contributed either unique expertise in document planning or extensive experience using the computer system, and second, the task of estimating, planning, developing consistent standards for so many manuals required a broader perspective and more carefully considered decisions than a single writer could manage alone. In this case, even though the writers had no idea which specific manuals they might write individually, they perceived the estimating and planning stage as a crucial moment in the project which required intensive shared decision-making responsibilities.

In addition to the two special examples mentioned above, writers and managers agree that coauthoring had several other benefits, including improving the quality of the final product, providing multiple perspectives on the topic and format, and relieving a single writer of the tedium and isolation writers suffer during long writing periods. When coauthoring, writers prefer to divide a manual into sections, with writers working independently to write their own sections, and then working together on the final drafts to ensure that the product looks as if it had been written by one person.

One manager indicated that the purpose and format of a document makes a difference: a reference manual can more easily be divided between multiple writers than a tutorial or users' manual since the reference manual is already subdivided into discrete sections independent from one another, but the tutorial relies on a cohesive developmental structure better handled by a single author. In every case, the writers agreed that cooperative writing and coauthoring require maturity and good planning and management skills so that individual responsibilities are well defined at the outset. In addition to adhering to the writing plan, writers must continually communicate with each other; otherwise, the collaborative experience falls apart.

WHAT DOES CLASSROOM COLLABORATION ENTAIL?

Writing for Computer Technology, an advanced technical writing course, was the basis for our case study of classroom collaboration. Although the course focuses on writing software documentation for novice computer users, it stresses such fundamentals as clear and concise technical prose, audience analysis, document planning, and document design. The twelve students in the course included two juniors and ten seniors: all completed the lower division prerequisite, Technical Writing, and ten completed the upper division course prerequisite Professional Writing in Business, Government, and Industry. Consequently our population of student subjects included relatively experienced writers, ten of whom had previous collaborative writing experience in the advanced class as well as word processing experience in other writing courses. For the collaborative assignment, the class divided into teams of four. Each team was responsible for documenting a different software system in the microcomputer classroom:

- the Xerox Ventura Desktop Publishing system: writing a tutorial for beginners;
- the WordPerfect word-processing program: rewriting a tutorial for beginners and a reference section covering advanced commands and features; and

- communications software: writing a users' manual for beginners who want to transfer files between computers and use electronic mail, the campus bulletin board, Writer's Workbench, and other programs available on the mainframe computers.

The assignment included a jointly written planning document, weekly status reports from individuals, two group progress reports, a collaborative oral report, a collaboratively written final document, and peer evaluations. (See Appendix A for a complete description of the "Collaborative Writing Assignment" for the course. While this assignment involved specific software available at the University of Wisconsin, the tasks are easily adaptable to other software packages.) Our case study data include pre- and post-semester surveys about students' writing and collaborative experience; audiotapes of group planning and collaborative writing sessions; audiotaped final interviews with individual students; and all written work produced by individuals and teams.

The classroom assignment used in this study asks students to "work collaboratively with other members of the class to produce a usable document for the microcomputer lab" and defines "collaborative writing" as "working together cooperatively with a team of individuals who are responsible for producing a final project" (see Appendix A). In other words, this assignment requires consensus in the production of the planning document, rough draft, group progress report, completed document, and the group's oral report. This consensus distinguishes this project as shared-document collaboration rather than as team work (or what other researchers have called "group work") in which the author may accept or reject peers' recommendations [12, pp. 72-73; 17, p. 55].

What was most interesting to us is how each group took a different path toward achieving consensus on the final document, and in doing so, managed to employ every form of team work and shared-document collaboration we identified as typical in private industry. Although the writing task was similar for each group, the group dynamics and collaborative processes varied within groups according to the relative skills and technical knowledge of group members, personality conflicts and alliances, and the lines of communication which developed within the groups. In each case, the groups' collaborative processes were shaped by two kinds of issues: first, how to use the individual skills of group members in the most productive way for the particular writing task, and second, the social and interpersonal dynamics that developed within the group. The following description sketches the kinds of collaboration employed by each writing team in the class.

Group A: Writing the Ventura Tutorial

The first group faced the difficult task of researching and mastering a complex desktop publishing system and writing a tutorial which would introduce a complete novice to the system's concepts and provide step-by-step instructions on how to use the system. To achieve this goal, they employed shared-document decision making throughout the project. They discussed and developed their initial planning document by consensus so that each person played a special role in the development of the project as a whole. The group leader led the group by facilitating communication, scheduling meetings, taking care of loose ends, but most importantly by steering the group through conflicts to successful resolutions. The content expert constantly advised the group on technical difficulties with the computer system, while the third group member developed graphics for each section of the document, and the fourth acted as stylistic and page layout editor. Additionally, the fourth member, who possessed less technical knowledge than other members, played the role of a test audience or novice user for the group as a whole.

The consensus approach also worked well in the group's research effort since each member got hands-on experience with the computer system independently, and then the group met as a whole to discuss and share their individual learning experiences as a basis for developing a unified tutorial approach and format. Similarly, each person wrote separate sections of the document and then the entire group met for collaborative editing sessions, which proved to be a time-consuming but effective method for shared decision making. When time constraints did not allow face-to-face collaborative editing sessions, group members exchanged copies of their drafts, edited them, and returned them for individual writers to revise. The group leader emphasized that "We did this continually and ended up with wonderfully revised drafts," while a second person said "I loved it. Writing alone with group follow-up works," and a third said "In this way, we were all able to contribute equally to the manual."

Even this group's oral report demonstrated an equal distribution of shared responsibility and knowledge. The completed document was a thirty-six page tutorial divided into eight distinct sections, and yet, stylistically and organizationally, it looked as though it had been written by a single author. Partly because a tutorial document requires developmental and organizational consistency throughout and partly because this group hit upon a workable group dynamic and document planning strategy early on, their group processes followed a successful pattern of shared-document collaboration at crucial decision-making, research, and editing levels.

Group B: Rewriting the WordPerfect Manual

The second group's task involved rewriting an existing manual for WordPerfect, the word-processing program used in the microcomputer lab. The task required creating a uniform style and improved page layout for the manual, adding new sections for advanced functions, revising unclear and incomplete sections, and improving the manual's readability and usability. The second group employed a mixed set of collaborative practices, including a limited form of shared decision making in the initial planning stages, coauthoring of individual sections, and peer editing. Collaboration as a whole occurred only in the beginning when the group met to develop the initial planning document. The group also devised a reader survey and shared in collating and evaluating the data.

Initially, the group set high goals for completely redesigning, rewriting, and expanding the fifty-page WordPerfect manual written by last year's class. However, they quickly felt overwhelmed by the task and had difficulty agreeing upon how to proceed. Consequently, they retrenched, lowered expectations, and divided into two subgroups of two people each. One subgroup thoroughly revised the tutorial section of the manual, while the other subgroup wrote a style guide for the whole manual and rewrote the reference section of the manual. Although each person initially claimed a participatory role in the larger group (group leader, editor, graphics coordinator, and content expert), the roles never materialized once the group divided into two separate subgroups. After the first two or three planning meetings, the two subgroups had little contact with each other except for peer editing exchanges and one occasion of group decision making, when the style sheet was reviewed, critiqued, and approved by everyone and then followed faithfully by both subgroups. In fact, the style sheet so successfully standardized document design, layout, and writing style, that the two subgroups were liberated from each other to perform independent work, and yet each managed to produce separate documents which fit transparently together in a single manual.

The first subgroup worked closely together as coauthors, completely rewriting the fifteen-page tutorial section of the manual word for word on the computer screen. One of the two coauthors described this process as follows: "I typed while she watched and we both critiqued, each contributing to every word and sentence; sometimes she'd suggest one word, then I'd suggest another, then simultaneously we'd both hit upon a third word, exactly right for that particular context." The coauthoring process was immensely time consuming and often frustrating, but the final rewrite completely transformed the old tutorial section into a superior document. Both these writers vigorously praised the coauthoring process as a means of achieving a higher quality document. The second subgroup of two writers coauthored the style sheet and group memoranda (the group progress report and instructions to outside reviewers) and wrote separate sections for the reference segment of the manual, frequently consulting each other for editorial and organizational advice.

Because of time constraints, the WordPerfect group never reconvened as a whole to combine their two efforts into a single finished document, and the two subgroups turned their work in separately (in different manila folders and on different computer disks). However, because of the stylistic unity of their final products, it was easy to splice the sections together into the new manual. Although this group never achieved full consensus as a four-person writing team, they successfully employed a combination of collaborative strategies (peer editing, labor intensive "divide and conquer" collaboration, and coauthoring) to fulfill their shared-document responsibilities for the course.

While the teacher intervened early on in helping the group set more manageable goals, he chose not to intervene when the group split into two subgroups because he felt the group would benefit from the experience of solving conflicts on their own and because the two subgroups worked effectively. While this form of collaboration does not fit our ideal, in the case of severe personality conflicts, it may be a realistic and productive solution. Later we discuss various modes of intervention teachers might use to handle difficulties with the collaborative process.

Group C: Writing the Communications Manual

The third group wrote the "Communications Manual" for the microcomputer lab. This involved writing a manual explaining how to communicate with the campus mainframe and use the software, such as electronic mail and Writer's Workbench, available only on the larger computers. Because of the technical difficulties involved, such as converting files and transferring them between machines, the group was initially overwhelmed by the amount of research involved. Therefore, they, like the WordPerfect group, used a labor intensive, divide-and-conquer strategy, which included a combination of collaborative strategies, such as minor coauthoring, some shared planning and document design decisions, but mostly individual authoring augmented by peer editing. Each group member wrote a separate section of the initial planning document; then the group as a whole met in an on-screen coauthoring session to merge their sections and revise the overall document.

In the initial group discussions, the team shared decisions about the audience, format, and overall organization of the manual. However, soon after they made the initial scope decisions, the communications group tended toward individual task division for several reasons. First, since the manual involved documenting several separate but related programs, they divided the manual into modular sections which could be worked on independently. Second, although the group identified individual roles (group leader, graphics coordinator, technical expert, editor), the roles materialized in an uneven manner, with the group leader scheduling meetings and keeping track of individuals' progress and the technical expert doing considerable research for the group as a whole, but without the graphics coordinator and editor fulfilling their roles. Third, the group's technical expert possessed so much more computer knowledge than the other group members that he dominated discussion, alienating the members who had less knowledge or interest in the technology. A hierarchy developed in which the more technically knowledgeable individuals tutored and led the less technically knowledgeable members, which diminished the peer group decision-making process, but efficiently disseminated information so that each member could write independently.

Up until the last week, the group exchanged drafts and commented on each other's work and occasionally reconvened to consult how everything would fit together. However, the least technical member of the group participated less and less as the project advanced, and by the end of the class, she merely turned in a disk copy of her section without showing it to the others. Ironically, at the same time the technical difficulty of learning the computer system tended to isolate group members, it also bound them together: this group regularly shared electronic or disk copies of their files and frequently used electronic mail to keep in touch and exchange ideas and technical information. So although the writing process appeared less collaborative than in the other groups, the technology itself provided communication tools which facilitated collaboration in a broader sense. Consequently, the communications group completed a forty-page manual which everyone felt proud of: one member said he was most proud of "the amount of information we've covered—three different computer systems and six different programs;" a second said "I like using the review process. It's easier to carry out a project when the group makes decisions instead of the writer alone being responsible for every decision. No single writer could possibly cover the bulk of information that we did in the same detail and amount of time that we did;" and the other two members were proud that they finished the job, emphasizing that at the beginning "we thought it was impossible." Even though the communications group appeared to be the least cohesive of the three groups, its members clearly perceived their accomplishment as a collaborative shared-document production.

HOW DOES THE COMPUTER FACILITATE CLASSROOM COLLABORATION?

The computer environment facilitated classroom collaboration in three important ways:

- The computer hardware, software, and procedures form the basic subject of research and a unified framework for developing the groups' awareness of audience needs.
- The word processor provides an ideal writing and revising tool for peer review and the constant rewriting required by the collaborative assignment.
- The computer classroom and its facilities provide a common medium for coauthoring, sharing and merging texts, and a common meeting place for group research and writing.

Computer Technology as Research Subject

Since the assignment required learning a technology to which everyone had immediate and equal access, the computer classroom formed a hands-on environment in which group members could experiment together and share knowledge with one another. Furthermore, as Timothy Weiss points out, "The word-processor and the business-writing student seem a perfect match. After all, business-writing students know that they will use microcomputers throughout their careers . . ." [20, p. 57]. This was especially true of our technical writing students, one of whom said in his post-semester survey: "I'd wanted to learn more about computers for a long time, and this was an opportunity to learn for and with other people, just like in the 'real world.' " Since the manuals created by the students in this class are used by students who take courses in the micro-computer classroom, the writing teams became heavily invested in their projects and frequently emphasized the goal of creating usable documentation. One group member asserted that "The most rewarding aspect of the group work was finishing a good, comprehensive set of instructions that *people can use*."

In each of our groups, the most seriously negotiated issue was the problem of satisfying the needs of the computer classroom's user community. John Trimbur, as quoted by Harvey Wiener in "Collaborative Learning in the Classroom," suggests that the "group's effort to reach consensus by their own authority" and the "intellectual negotiation that underwrites consensus . . . promotes a kind of social pressure" that leads students to take their task more seriously and "fight for their ideas, or modify them in the light of others' ideas" [17, p. 54]. As experienced users of the hardware and software themselves, students varied in the degree to which they empathized with the novice user's perspective. Those who identified with the novice user pressured the more technically advanced group members to accept or negotiate their concept of the reader. For example, Rod, the technical expert in the Communications group, repeatedly asserted that exact technical terminology and a detailed description of the system was what students needed most. However, the other members resisted this idea, at first joking about "technical elitism," and later seriously arguing that Rod was wrong and that beginning users needed simple language and easy-to-follow instructions. Eventually, the opinion of the other three group members prevailed, and Rod reluctantly agreed to replace the technical terminology in the manual with simpler language. Similarly, in the other two groups, the task of producing usable documentation for student users created social pressure and a serious commitment within the group to arrive at a consensus about the audience needs and the goals of the collaborative task.

Computer as Collaborative Writing Tool

As the primary writing tool, the computer facilitated the collaborative effort in several ways. In the post-semester surveys, all class members agreed that word processing made group writing easier, especially during the revising and editing stages. Survey comments repeatedly emphasized that the computer eliminated the arduous task of retyping, that changes could be made quickly, that the screen was "less messy" than paper, that document formats could be easily changed and experimented with, and that the drafting process went faster. In general, the word processor made students much more willing to rewrite and respond substantively to peer commentary. In an article on student writing and word processing, Diane Pelkus Balestri identifies this ability to construct, test, change, and reshape text on the computer screen as the major advantage of "softcopy" over "hardcopy": "Softcopy is malleable . . . it can be expanded and contracted, split and joined, parts reserved, discarded and easily inserted . . . text is easily transformed in its appearance and content, with different fonts and other graphical embellishments. It enables a writer to consider every aspect of the presentation of the text to an audience . . . it allows the writer the unique freedom to engage in a wide range of writing activities" [21, p. 17]. "Softcopy," then, not only makes it easier for students to incorporate and try out each other's ideas, but it also makes it easier for them to coauthor and coedit shared documents on the computer screen. Kate and Claire, who coauthored the WordPerfect tutorial, spent hours trying out different phrases and formats on the screen: "One minute there was all this ugly text [on the screen], then we'd get a new idea, and 'swoosh' the old text was gone and the new one looked so much better. It was fun, kind of like playing a game." Some students acknowledged that they spent more time than expected when working on the computer because they composed and revised more extensively on the computer and experimented with formats. Even so, all agreed that collaboration in the computer classroom enhanced the quality of the final document. One lesson for teachers which emerges from our study is the need to provide in-class time for both collaborative and computer work.

Computer Classroom as Collaborative Environment

Finally, the computer classroom itself forms an ideal atmosphere for collaborative learning and writing activities. Student focus shifts away from the teacher at the front of the room toward the activity of writing and manipulating text on the screen. One researcher, Timothy Weiss, states "it is difficult to lecture in the computer classroom—to get my students' attention I must order them away from their computers!" [20, p. 63]. Furthermore, students frequently turn to each other for help or advice, with more technically advanced students helping out and even tutoring those less familiar with computers. The public nature of the computer screen combines with this exchange of knowledge so that students become more willing to share their writing and ideas. Several comments from the students' post-semester surveys indicate how the computer classroom enhanced collaboration:

It became a central meeting place outside of class, not only for the group, but for other users whom we interviewed and got information from.
We had all our meetings there because immediate access to the microcomputers and other lab resources meant that if we had questions, we could show each other on the computer.

> We kept copies of each other's files on disk (much easier than carrying around paper copy) and could just print or edit a copy whenever we wanted ... also, if one member understood part of the system that the rest of us were puzzling over, she could just give us a quick demonstration to clarify the problem.
>
> We were able to leave messages on e-mail (electronic mail) to let other group members know what we discovered, figured out, etc.—it made keeping in touch easier when dealing with busy schedules.

The exchanges by electronic mail tended to be informal and chatty, increasing the back and forth, conversational flow of written communications. Once the Communications group taught the rest of the class to use the system, nearly all of the weekly status reports were sent by electronic mail, which resulted in more frequent and more informal written exchanges with the instructor—over 150 electronic mail messages were exchanged between the students and the teacher alone—and probably many more exchanges occurred among the students themselves. Consequently, the computer classroom fostered a ready exchange of texts, disks, ideas, and knowledge among peers and helped disseminate the pedagogical authority of the teacher into the collaborative activities of the students, who became more productive managers of and authorities over their own writing projects.

WHAT DEVELOPMENTAL WRITING AND SOCIAL ISSUES EMERGE IN CLASSROOM COLLABORATIONS?

The difference between the issues which emerge from classroom collaboration and those which emerge from workplace collaboration results from the difference in the aim of collaboration within each institutional setting. While the chief aim of workplace collaboration is productivity, the chief aim of classroom collaboration is student learning. Nonetheless, the lessons from corporate and classroom collaborations are mutually instructive.

Our case study of classroom collaborations reveals two learning-related issues: the first concerns the effect of collaboration on individual student's writing development; the second concerns the development of what Morgan et al. call "groupness," that is, the development of a group identity and strategies for working together [13].

Collaborative Assignments and Individual Writing Development

In addition to confidence and competence, writing maturity, particularly the sort expected of technical writers, requires intellectual maturity as well. In technical writing literature, this kind of maturity is often described as "egolessness." This does not mean that technical writers have no investment or pride in their writing nor that they automatically defer judgments about their writing to others. Rather it means that they can detach themselves from their writing, accept and evaluate the criticisms and suggestions of others, and incorporate both their own and others' views in planning, revising, and editing their texts.

In *Forms of Intellectual and Ethical Development in the College Years*, William Perry contrasts intellectual immaturity with intellectual maturity in the following terms. Immaturity is characterized by dogmatic, either/or thinking; ego-centricity or an inability to recognize and value alternate perspectives; and belief in absolute truth which is transmitted by various kinds of absolute authorities. Maturity, conversely, is characterized by nondogmatic, flexible thinking; recognition of the value of multiple perspectives; and acknowledgment of the relativity of truth and the contingent nature of authority [22]. Though other scholars like Carol Gilligan, Mary Field Belenky et al., and Patricia Bizzell question the gender and class bias of Perry's findings, they generally view the movement from simplistic to more complex thinking as developmental in nature [23-25]. Though it is beyond the scope of this study, an examination of gender differences in classroom and workplace collaborations could offer new insights into the relationship between intellectual development and group dynamics.

The connection between the ego-lessness required in technical writing and intellectual maturity seems clear. Flexibility, receptivity to alternate views, and awareness of multiple and sometimes conflicting authorities are essential to the work of technical writers who regularly consider the needs of various readers; elicit and incorporate information from technical experts; juggle feedback from peers and managers; and make informed judgments based on a recognition of their role in the larger corporate structure.

The advanced technical writing students in our case study exhibited ego-lessness to varying degrees. An early indicator of students' developmental levels was their responses on the initial survey. Not surprisingly, the students with the strongest writing background expressed a positive attitude toward such collaborative activities as peer critiquing: "I'm pretty confident about my writing and I've been critiqued almost to death in writing classes, so I've lost my 'ego,'" commented one such writer. Less able or less experienced writers, on the other hand, expressed anxiety or mixed feelings about the prospect of peer critiques: "Writing is very personal, even tech writing. I want criticism to improve, but do not want my ego shattered," wrote one less experienced writer. Paradoxically, "ego-less" writing requires a healthy writing ego.

Since collaborative assignments involve regular feedback on individual participant's writing and accommodation of various points of view, these assignments should, in theory, promote the kind of writing maturity we have been describing. In some instances, we found this was so, while in other instances we observed no apparent growth during the semester—however, developmental theory suggests that a lag often occurs between learning and behavioral change which means that growth may have occurred but was not yet observable. To illustrate how intellectual maturity or ego-lessness affects and is affected by the collaborative process, we offer two contrasting cases.

Although Rod was technically knowledgeable and competent as a solo writer, he lacked the intellectual maturity necessary for ego-less writing. This intellectual immaturity was manifest both in Rod's social interaction with his group (Group C—Communications group) and in his response to peer critiques. In group meetings, Rod dominated group discussion, seemingly unaware that his detailed technical explanations were confusing and boring to other group members. Although Group C managed to produce a successful document, the group process suffered as a result of Rod's dominance and one group member withdrew almost completely. On the post-semester survey, Rod wryly comments on his attitude toward the opinions of others: "Getting them to see what is perfectly obvious to me was the most frustrating part of the collaborative process. (That's sort of a joke.)" Although the parenthetical aside may point toward a more mature perspective which recognizes his part in group problems, his behavior indicated that either his assertion was ironic and this was no joke or that he was unable to enact this more mature perspective. Because Rod felt there was a right and wrong way to do things and his way was the right one, he failed to see the value in collaboration and merely found it bothersome. In his exit interview, he summarized his distaste for collaborative work: "I wouldn't choose collaborative writing projects over individual ones unless I could choose the collaborators. I prefer having total responsibility for the finished output." Clearly, Rod prefers to feel in complete control and is, as yet, reluctant to let go of his ego-investment in a piece of writing long enough to benefit from the critical responses and suggestions of peers. Only once did he grudgingly concede to group opinion, but only after he had tried and failed to enlist his teacher's support for his position. (He had used the phrase "terminal emulation" in his documentation, and they insisted upon the less technical phrase "communicating with another computer.") Perhaps the strongest evidence of his continuing uneasiness about the ideas and work of his peers was his request to take an independent study the next semester in which he would rewrite the entire manual produced by his group.

Happily, other class members made observable gains in maturity as a result of the collaborative project; perhaps the most dramatic example was Carol. Initially one of the least experienced writers in the class and one of the most tentative about receiving peer critiques, Carol offered the following self-assessment during her exit interview: "With teachers' evaluations I always felt like I had to conform to their suggestions, but I felt different about my group's evaluations. They were trying to help me with *our* project; they made me feel that I wasn't the only one responsible for the quality of the writing. Strangely enough, this made me feel better about my own writing." In Carol's case, the shared responsibility was liberating. She wrote more confidently, knowing that her group would offer constructive criticism, not to find fault, but to make her writing and the overall document better. The collaborative team offered a social context in which Carol could shift the authority for judging the effectiveness of her writing from an external single authority figure to a peer group in which she was a full participant.

Carol's gains in writing maturity were manifested in several ways. The writing she produced for the collaborative document was stronger in style and content than the writing she produced on her own earlier in the semester. In addition, Carol took a major role in the group's oral presentation. According to her teacher, "The self-assured, articulate Carol who gave the oral presentation hardly seemed like the same person who, at the beginning of the term, prefaced every question with 'I know this is a dumb question, but . . . .' " Final testimony to her new sense of herself as a writer was her decision at the end of the course to apply for a technical writing internship for the fall, a decision she claimed she would not have made had it not been for the collaborative writing experience. After successfully completing her internship, Carol was hired as a full-time technical writer in a local corporation.

The transitional "interpretive community" offered by the peers in her writing group allowed Carol to move from an insecure writer who looked to the teacher for all judgments to a more confident writer who looked to herself and her peers for judgment. Rather than view Carol's inexperience with computers as a disadvantage, her group designated her spokesman for the user, turning what could have been a weakness into a strength, a source of authority. Convinced that she had something to offer, her confidence in her opinions and writing contributions grew.

Several lessons for the technical writing teacher emerge from these two cases. For one, teachers need to monitor individual development by requiring class members to keep journal accounts of their group experiences. This way a teacher may intervene with written responses, conferences, or activities aimed at helping an individual writer whose immaturity is impeding the group process or whose growth is being impeded by others' immaturity. Goldstein and Malone offer both practical advice and testimony to the effectiveness of this strategy: "Through reading the journal, the instructor sees the group's world through the eyes of each member and can provide appropriate feedback to individuals . . . without stepping in and taking over" [26, pp. 113-131]. Moreover, teachers may wish to facilitate intellectual growth through group problem-solving exercises such as those recommended by Kenneth Bruffee and Karen Spear [27, pp. 8-19; 28, pp. 159-162]. The purpose of these exercises is not so much to prevent group conflict as to give students a framework for discussing the problems which inevitably occur in the group process.

We caution teachers, however, against too much intervention. The teacher can do only so much to encourage individual development in the peer group process. Direct intervention of the sort a technical writing manager might undertake may be pedagogically unwise since it can interfere with the peer groups' need to identify and solve group problems as part of their learning process.

Collaborative Assignments and Group Development

As the earlier sketches indicate, the forms of collaboration used by our case study groups varied considerably. The factors that shaped the collaborative process include the following:

- the use of the computer lab environment to foster group interaction;
- the dynamics of the first planning meeting, frequency of subsequent meetings, and the ability of the group to integrate all members; and,
- the variety and quality of the group's collaborative conversations.

Group Development and the Computer Classroom Environment

The ability of each group to collaborate in one way or another and to achieve some sense of "groupness" was in large part due to the computer lab environment. For everyone it was common ground: a place to meet formally and informally, a place to communicate and share texts electronically, and a laboratory for research on the software they were documenting. For the Ventura group, which met frequently and consulted with one another every step of the way, the lab was a place to conduct face-to-face meetings. For the other two groups, the lab provided a means of communicating electronically, without face-to-face meetings. Finally, and not least important, the lab was a place to come into contact with the users for whom each group was preparing its document. Thus the lab provided a larger "interpretive" community within which each group attempted to identify its smaller "interpretive" community.

Group Development and the Evolution of Group Dynamics

Although each group discovered methods of collaboration which allowed it to complete the assigned task successfully, the Ventura group's collaboration involved the most group interaction and resulted in the most unified document. Therefore, this group's collaboration merits further analysis and offers a point of reference for discussing the other groups. The elements which contributed to the effective group interaction of the Ventura group were evident in their first planning meeting. Even though they, like the other groups, felt overwhelmed by their task, they made three important procedural decisions at this meeting: they agreed that each person would experiment with the system on her own and take notes; they agreed to meet outside of class on a weekly basis; and they agreed on a minimal list of features to be documented. The decision to experiment individually resulted in an initial set of common experiences; the decision to meet weekly fostered a sense of group identity as well as effective communication;

and the decision to document only a minimal list of features provided an initial sense of direction. Furthermore, by postponing decisions about divisions of labor, they avoided early disagreements over turf, and by putting off premature decisions about scope, they avoided arguments about goals.

By contrast, the first meetings of the other two groups proved frustrating, setting in motion personality conflicts which were never resolved. In the Communications group, Rod dominated the conversation, explaining the intricacies of the software and generally directing group planning. In the WordPerfect group, a personality clash emerged between the group's leader, Ann, and another member, Kate, making their early conflicts over procedure and scope vexing to all. Since neither of these two groups chose to meet regularly outside of class, it seems that the success of the first meeting was crucial in laying the groundwork for subsequent collaborative activities.

Another reason for the Ventura group's cohesion was its ability to integrate all members. As mentioned earlier, Carol, the least technically knowledgeable member, assumed the role of the novice user. "I represented the user," she said. "I would say, 'Wait a minute. I don't understand and if I don't understand probably the novice user won't either.'" Her inexperience thus provided her with a role that was valued by the group. In the other two groups, however, differences in writing ability, technical background, and personality diminished group cohesion. For example, Pam, the least experienced writer in the Communications group, felt she had nothing to offer and increasingly became disengaged from the group effort. In the WordPerfect group, the four members split into two teams because of personality conflicts, and each team worked out its own method of collaboration.

Group Development and
Forms of Collaborative Talk

Finally, we found that the forms of collaborative talk revealed a great deal about the dynamics of the groups. From the beginning, the tapes of Ventura group meetings could be characterized as conversations. Though Carol's voice is heard less frequently and sounds more tentative in the first tape, by the second tape she is a full participant in the conversation. By contrast, the talk which occurred in the other two groups was rarely a conversation which involved the whole group. By Tape #2, the WordPerfect group interaction had broken into two conversations which occur simultaneously. These two competing conversations—one between Ann and Julie, the other between Claire and Kate—represented the two sub-groups which emerged. As we listened to the tape, we could actually hear the alliances forming. As for the Communications group, their interaction often involved a monologue with Rod giving long-winded technical explanations to the others. Like the other two patterns of verbal interaction, this pattern matches the key dynamic at work in the group. A conversation among four marked the interaction of the group that worked as a cohesive team; two separate dialogues marked the interaction of the group that broke into two pairs of collaborators; and monologue marked the interaction of the group that worked as four loosely related individuals.

Listening to these collaborative conversations suggested distinct types of group talk. While we found no other study of collaborative writing which offers categories of group talk suitable to our study, Mary Ann Janda's 1988 dissertation, "Talk Into Writing: How Collaborative Writing Works," offers an insightful analysis of the relationship between talk and collaborative writing in a variety of contexts [29]. Janda identifies the collaborative writing process through such activities as forming group identity, exploring the subject, and producing the text [29, pp. 76-77]. Whereas Janda examines the dimensions and stages of collaboration, we explore how group talk shapes group interaction. To do so, we developed the following four categories for distinguishing group talk: procedural talk, substantive talk, writing talk, and social talk. Procedural talk involves division of labor, setting of goals and ordering priorities, revising previous plans, and making arrangements for meetings and exchange of drafts. Substantive talk is subject-related and includes sharing technical information, explaining processes, asking questions, and solving problems. Writing talk includes discussions of the writing process, discussions of style, commenting and editing drafts, and coauthoring, while social talk entails exchanges about outside activities, friendly bantering, and community-building talk.

Comments like the following illustrate procedural talk: "First we better each try and play around with this system;" "Let's keep a joint computer disk with all of our drafts on it in the lab, so we can all go in and look at them when we're in the lab;" "One thing we have to do is work out ways to keep users on track." Throughout the collaborative process, procedural talk was a necessary part of each group session; however, such talk tended to diminish as the groups evolved and established methods of working together. When group members could not agree on procedures and took undue time trying to reach consensus on these matters, the group process broke down and everybody became discouraged. The WordPerfect group was a case in point. Their time-consuming haggling over procedures, often a mask for personality conflicts, hindered group progress and damaged group morale. The solution they eventually settled on was to commission two members to draft a style sheet which then became the key procedural guide.

Substantive talk—"Did you know that when you use Ventura, you should edit and spell-check everything first? That's another reason why we should recommend doing early drafts on WordPerfect"—increasingly became the focus of group conversations as the work progressed. Lengthy discussions of substance in the first meeting, in fact, worked against full participation, for such discussions were dominated by the one or two people who were more technically knowledgeable than the others. This problem was particularly evident in the tape of the Communications group's first session. Our study suggests that the degree to which substantive talk involves all members and becomes a dialectical process determines to a large extent whether or not the group acts as a team or breaks into splinter sub-groups or a collection of individuals, each working more or less independently. The mark of a mature and cohesive group is one that can use conflicts over substance as a way to clarify and enrich the writing produced.

Of particular interest to us was the writing talk which ranged from discussions of the writing process—"I'm a real good proofreader of everybody's text but my own"—to discussions of computer terminology—"What else can you say for *retrieve*? *Get*? *Fetch*?" Our general observation was that the most cohesive group, the Ventura group, interspersed writing talk into most group sessions, particularly as the project progressed. The other two groups discussed writing per se very little, preferring instead to comment on one another's writing by using electronic mail or exchanging disks.

A notable sub-category of writing talk was the coauthoring talk which occurred between Claire and Kate in the WordPerfect group. The ninety-minute tape of one coauthoring session offers insight into this particular mode of composing. Claire sat at the keyboard with Kate beside her. The give and take dialogue evident in these composing conversations is similar to what Vygotsky and others describe as the "inner dialogue" which occurs when a single author writes alone [30]. One person would begin a sentence which the other would complete or one person would generate a sentence which the other would revise. As Bruffee suggests, this kind of peer conversation elaborates thought and results in both intellectual growth and a more effective written text [31]. At times Claire and Kate's conversation became so blended that they communicated in code. "Kate would reach over and move the cursor down and she wouldn't even have to say 'Bold it.' I'd just know that's what she meant," reported Claire. The computer as composing tool contributed significantly to the coauthoring fluency which developed between Kate and Claire.

A final form of talk which occurred was social talk. Although no group became side-tracked with extensive social talk—at least not while they were being taped—we found that some social talk benefited the group process. As one might expect, the Ventura group which met the most frequently engaged in more social talk than the other two groups, exchanging personal information about how group members managed to get work done with small children around and so on. Interesting to observe, however, was the way in which social talk blended with substantive talk for this group as they shared anecdotes about struggles with the system. Similarly, for the coauthors, Claire and Kate, social talk—teasing, gossip, jokes—was interspersed with composing talk, substantive and procedural talk, cementing the bond that developed from their collaboration.

The pedagogical implications of our study suggest that students engaged in collaborative projects might benefit from concepts of group dynamics and conflict management. These concepts could be introduced experientially through classroom activities or through observation and analysis of videotaped collaborative sessions. But as with the matter of individual development, part of the learning involves allowing the groups to develop their own sense of "groupness" and modes of collaboration.

For us, the study of in-process classroom collaborations provoked questions more than it provided answers. Do certain forms of collaborative talk occur in all types of collaboration or are these forms context specific? What strategies do groups use to resolve conflict? When is instructor intervention appropriate? What similar interactions occur among workplace collaborators and managers? Clearly, further study of in-process collaborations are needed; such studies offer researchers an opportunity to observe interpretive communities in the making.

TOWARD A THEORY OF COLLABORATIVE DISCOURSE FOR TECHNICAL WRITERS

Throughout the literature on technical writing research and pedagogy, there is a tacit assumption that the technical writing classroom should strive to simulate the practices and situational contexts of the workplace [32, 33]. We agree that writing pedagogy should comprehend what experienced writers actually do, but we would like to challenge the assumption that the classroom should always imitate workplace practices and goals. Workplace and academic settings represent related but distinctly different "interpretive communities," or established communities of writers and readers who construct knowledge out of shared assumptions, conventions, and discursive practices. The technical writing classroom shares many rhetorical conventions and writing procedures with the working world, but we cannot assume that an unchanging, universal set of rhetorical principles binds the academic writer to the same uses and interpretations of those conventions as the workplace writer. John Trimbur points out that each community of discourse derives its specific formats and conventions from a disparate and historically determined set of practices:

> . . . the genres and formal features of writing—whether a case study, belletristic essay, a business report, or a scholarly article—are neither expressions of underlying rhetorical principles nor manifestations of deeper cognitive structures. Instead, they are discursive practices, historically derived through a process of group ratification, that bind interpretive communities together [34, p. 100].

Historically, the writing practices of the workplace have been shaped by productivity and management objectives distinctly different from the educational objectives of the academic setting. For example, we found that writers of software documentation rarely practice coauthoring because it hinders productivity. Similarly, team work and shared-document collaboration in the workplace serve primarily to ensure the accurate reproduction of meaning and to streamline the writing process for greater productivity, such as when multiple reviewers check the accuracy of content, peer editors help develop audience awareness and stylistic clarity, and group standards enable writers to invoke standardized templates and writing formats for consistent, effective, and rapid document design. In the writing classroom, on the other hand, we found that peer collaborations foster the traditional educational goals of intellectual growth. Even when collaboration appeared to result in a less productive writing experience, it clearly fostered a richer learning environment for development at both individual and group levels. Although students found coauthoring and other collaborations time-consuming and frustrating, they described these same collaborations as rewarding and worthwhile learning experiences. Industry allows little room for such developmental experience: professionals must already possess maturity as writers and communicators, especially when collaboration combines writing deadlines with office politics and business priorities. Although students should know about such conditions, the writing classroom could never completely simulate them and should not attempt to.

The research of Glenn Broadhead and Richard Freed in *The Variables of Composition* confirms this distinction between workplace and academic writing communities. Their detailed study of writers in an international consulting firm showed that experienced writers use "highly staged" writing strategies and a relatively linear writing process which relies on "stored representations— appropriate notions about purpose, audience, line of thought, or tone," stock sections, and boiler-plating to speed up drafting and eliminate the need for incubation, exploration, and conceptual and organizational revision [35, p. 131]. Concurring with the findings of Jack Selzer's study of a single engineer, Broadhead and Freed conclude that workplace writing differs substantially from the non-linear, "recursive-oriented" models of writing that composition researchers such as Nancy Sommers and Flower and Hayes stress as important discovery and fluency-building tools for the cognitive development of student writers [33]. This leaves advanced composition classes, such as business and technical writing, in a unique position. On the one hand, we face the task of developing the writing processes and cognitive awareness of students fresh out of composition classes, but on the other hand, we must teach them discursive writing practices, including forms, formats, and schemata specifically developed by industry to leap over problems of cognitive development and focus directly on the problem of the rapid production and reproduction of reader-based texts.

Clearly we cannot simply import or imitate the "stored representations" and "highly staged" writing strategies of a single corporate ethos and expect our students to be adequately prepared for the immense variety of situational contexts which change radically from company to company. At the same time, technical writing teachers should avoid retreating into comfortable academic models of discourse that oversimplify the complexities and variety of discourse conventions found in the workplace. Group work and peer collaboration diversify and disrupt the homogeneous, single-author model of writing in traditional academic classrooms by exposing students to a variety of collaborative practices and problems which help them learn what it means to make the transition from one community of discourse to another. As John Trimbur suggests, peer feedback between student writers enables them to act as translators between discourse communities, and hence creates the possibility of transition:

> By emphasizing the social activity of writing, collaborative learning can play a significant role in helping students make the transition from one community to another, from one discourse to another, from one identity to another. Collaborative learning can help students generate a transitional language to bridge the cultural gap and acquire fluency in academic conversation [34, p. 101].

Although Trimbur is talking about initiating students into academic discourse at the beginning of their academic careers, we feel that students leaving the academy can also benefit from technical writing courses which help them make the transition from academic discourse communities to the various discourse communities of the workplace. Collaborative learning, then, emphasizes the social contexts for writing, and thereby offers both a theoretical perspective for bridging the gap between academia and the workplace as well as a powerful set of classroom practices which foster student writing and social development.

Furthermore, collaborative writing works well in the technical writing classroom not simply because it reproduces the actual writing practices of professional technical writers and simulates workplace environments, but because it also teaches students how interpretive communities are formed and sustained. By forcing students to examine their own and others' assumptions about writing, collaborative writing assignments ask them to create, if only for a brief period of one semester, a shared set of conventions for producing and reproducing meaning in a precise and consistent way. For example, when the WordPerfect writing group discovered that they could not develop the social rapport required for close collaborative work, they constructed a style guide that bound each member of the group to an established set of writing conventions, ensuring consistency in their final group product. The WordPerfect group then reproduced their style guide for the other groups in the class who used it to settle their own disputes about format. This kind of learning experience teaches students how to recognize and value the highly staged writing strategies and stored representations used in differing workplace communities. When these students encounter the multitude of standards and established collaborative procedures in government, business, or industry, they may better understand how to adapt their writing skills to the differing demands of those communities.

Finally, collaboration itself suggests a model of technical discourse in which written texts can be understood not so much as products constrained by fixed rules and conventions, but more as interpretive events constructed by communities of readers and writers out of their shared assumptions and on-going negotiations over discursive practices. As Harvey Wiener suggests, collaborative learning is much messier in practice than in theory [17, p. 27] ; in the classroom, we may develop collaborative models as learning devices, whereas in the workplace, collaboration must constantly adjust such models to the varying demands of productivity. Even so, we can imagine a beneficial exchange between workplace and academic communities as based on collaborative dialogue rather than an imitation of practices. Such a dialogue might exchange knowledge and resources between communities without confusing their distinct institutional goals (see Figure 4). The diagram in Figure 4 expands on the collaborative model developed in Figures 1 through 3. Rather than representing "the review teams" as isolated readers, however, this diagram represents them as circles of "Student & Instructor Readers" and "Professional Readers" (right-hand group of circles within each dotted-line box) and as mirror images of the shared-document collaborators and decision makers in the circles of "Student Writers as Peer Collaborators" and "Professional Writers" (left-hand group of circles within each dotted-line box). Similarly, even though the writing conventions, reading habits, and institutional goals of academic communities may be distinct from workplace communities, Figure 4 emphasizes how the collaborative processes within each interpretive community resemble one another. Such a close structural resemblance may seem artificial for many forms of academic writing, but as our research reveals, for technical writing classes, the adaption of workplace collaborations to the classroom can be highly effective. The heavy-dotted-line boxes in Figure 4 represent the differing institutionalized structures of academic and workplace communities, each working within the permeable boundaries of its own established conventions and institutional goals, with "Developmental Goals" dominating the learning environment of the academy and "Productivity Goals" dominating the corporate environment of the workplace. As the curved arrows suggest, however, both communities are shaped by the collaborative nature of writing itself. Just as the cycle of reader response, peer exchange, and writer revision generates a dialogic construction and reconstruction of ever changing meanings and interpretations within a given interpretive community, a dialogic exchange between interpretive communities might provide an opportunity for growth and change within each community. Ideally, the academic community should be able to provide professional writers with a better understanding of how and why specific rhetorical principles work in specific writing situations, while the workplace can provide academia with experiential conventions and models for writing projects and assignments which help students understand the operation of interpretive communities whose assumptions and goals diverge from those of academic discourse.

But simply introducing collaborative writing assignments to the classroom is not enough, especially when students have been schooled to compete through

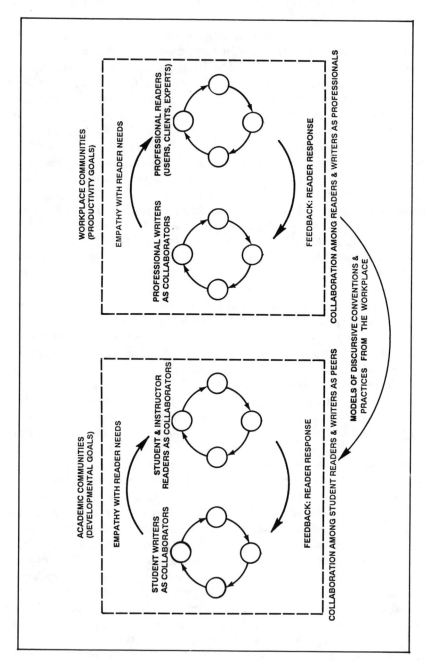

Figure 4. A collaborative model for communities of technical discourse.

individual writing performance rather than to cooperate in group settings. The fifteen-week classroom experience can't provide the same rich resources for collaboration that industry often provides, and teachers should not expect all students to quickly master the interpersonal and group skills required for successful collaborative writing.

Our research shows that the computer-assisted classroom can become both a collaborative writing tool and a collaborative learning environment which compensates for academia's lack of institutional resources—stored representations, style guides, management communication tools, document design standards, boiler-plate conventions, etc.—resources that normally sustain collaborative writing in industry settings. Two of the collaborative teams in our study suffered interpersonal conflicts that hindered group development. However, the computer technology acted as a kind of safety net which enabled the two groups to complete shared-document goals by sharing disks and document design formats, rapidly editing and merging files together, and communicating through electronic mail. The same technology that many professional writers take for granted as a foundation for successful collaboration also proved successful in the technical writing classroom—especially when the technology was fully integrated into the collaborative process.

Furthermore, the computer classroom helped create a nonhierarchical social context for an interpretive community of student readers and writers by encouraging a more public atmosphere for writing in which texts were created, easily viewed on the computer screen, revised and immediately printed out for group consumption, and shared as on-going events susceptible to constant change and individual as well as group participation. In this particular writing assignment, the peer identification with the student users, who frequented the computer classroom during open hours, cemented the collaborative bond between readers and writers so that participation in an interpretive community became a discursive reality which evoked a greater commitment from the students than academic discourse normally engenders.

Adapting collaborative learning assignments to the advantages of the computer classroom should liberate the energies of teachers and students for solving the difficult problems of group development. But even more importantly, the computer classroom also provides technical writing students and teachers with an arena for experimenting in the creation of interpretive communities, for understanding the exchange of rhetorical practices that emerge in such communities, and for identifying what it means to move from one discourse community to another.

APPENDIX A: CLASSROOM ASSIGNMENT

Collaborative Writing Assignment: Writing for Computer Technology

In this assignment, you will work collaboratively with other members of the class to produce a usable document for the English department microcomputer lab. "Collaborative writing" means working together cooperatively with a team of individuals who are responsible for producing a final project. You must fulfill at least two roles on the writing team: as the writer of a specific section of the document and as a specialist in one or more areas of expertise concerning the project. You are responsible for both roles. In addition to the documentation you produce, you must also learn to coordinate your individual effort, knowledge, schedule, and work habits with those of the other members of the group. This requires courtesy, thoughtful communication, and dependability on your part.

You will be given two grades on this collaborative writing assignment: the first grade will be the grade given to the project as a whole and the second grade will be the grade given to your individual performance as part of the writing group. The second grade will be an average of several grades: each team member will assign a grade to your work and those grades will be averaged with a grade assigned by the instructor to your role in the project work.

Each member of the team must write a specific section of the project—some members may write more than others depending on their role. Also various roles may overlap or be shared, depending on the needs of the specific project and team member skills. Members of each group may take on two or more of the following roles:

1. Writer: everyone will be responsible for writing a specific part of the project.
2. Group Leader: responsible for coordinating the rest of the team, organizing the documentation plan and the schedule—especially for group meetings, initial planning, and picking up loose ends.
3. Editor: responsible for editing, proofreading, and standardizing the writing style of the final document, and for guiding other writers in using a standard style, conventions, and format.
4. Graphics and Production Manager: responsible for document design, illustrations, layout, format, and printing of final document.
5. Content or Technical Expert: responsible for in-depth research on technical topics, assisting other team members with technical problems, and testing the final document for accuracy. Some projects may require more than one content or technical expert, especially when considerable research is involved.

As your instructor, I will act as your "manager" and divide the class into three writing groups. I will also act as final arbiter and advisor in the case of any problems or difficulties that arise. You, however, must meet as a group and decide which roles each of you will fulfill on the team and which sections of the project each of you will write. I will consult with each group before you finalize your individual roles on the writing team. Your group is responsible for producing the following:

A. An initial planning document, including a scope definition, an audience analysis, a tentative outline, a tentative schedule for completion of the project, and a clear list of specific tasks assigned to each group member.
B. A rough draft of the project which includes written sections, examples of the final document design, illustrations, and other visual aids.

C. A progress report on current project status, revisions, significant design changes, and a revised project schedule.

D. The completed document.

E. A group oral report and hands-on demonstration when appropriate.

F. A grade and brief (one or two sentence) evaluation for each member of your team.

G. A weekly status report submitted to me every Monday and copied to your group leader.

Sections A., B., and D. described above will be read and commented on by myself, technical advisors for each project, and the rest of the class.

Each group will be assigned one of the following topics by me:

- VENTURA GROUP: How to use the Ventura desktop publishing system; this project involves writing a section for the WordPerfect manual and teaching the other members of the class how to use Ventura to publish their projects and resumes.

- COMMUNICATIONS GROUP: How to use WordPerfect files with other systems, including: the "Punctuation and Style" program, the Kermit communications program, electronic mail, and "Writers Workbench" text analysis program. This project involves writing a section on your topic for the WordPerfect manual and teaching other members of the class to use these resources.

- WORDPERFECT MANUAL GROUP: Writing new sections for and re-designing the WordPerfect manual, which involves deciding what should be included in the WordPerfect manual, writing new sections, and rewriting and restructuring sections to accommodate the new sections, including those written by the other two groups. You will have to work with the Ventura group to redesign the manual according to features available on Ventura. Your planning document must prioritize potential changes to the WordPerfect manual.

Initial Group Status Report:

Tonight you will meet in your groups and each present your background, discuss your work so far in the course, and what you're most interested in doing for this project. By the end of class, your group must write a status report consisting of one paragraph on the background of the group and what roles specific people are interested in and a second paragraph on questions, problems, concerns, or bright ideas you may have for this project. One person should act as a recorder for the group during your discussion and take notes for the status report.

REFERENCES

1. M. Mullins, Technical Writer, First Wisconsin National Bank, Personal Interview, Milwaukee, Wisconsin, May 24, 1988.
2. L. Faigley and T. Miller, What We Learn from Writing on the Job, *College English, 44*, pp. 557-569, 1982.

3. P. Anderson, What Survey Research Tells Us about Writing at Work, in *Writing in Nonacademic Settings*, L. Odell and D. Goswami (eds.), The Guilford Press, New York, pp. 3–83, 1985.

4. A. A. Lunsford and L. S. Ede, Why Write . . . Together: A Research Update, *Rhetoric Review*, 5:1, pp. 71–77, 1986.

5. L. S. Ede and A. A. Lunsford, Collaborative Learning: Lessons from the World of Work, *Writing Programs Administrator*, 9:3, pp. 17–26, 1986.

6. W. Wheeler, Word Processing for the Technical Writer: A Case Study, in *Word Processing for Technical Writers*, Robert Krull (ed.), Baywood Publishing Company, Inc., Amityville, New York, 1988.

7. J. Forman, Computer-Mediated Group Writing in the Workplace, *Computers and Composition*, 5:1, pp. 19–30, 1987.

8. W. Van Pelt, Documentation—The Missing Link in Computer Technology, *Educational Computing*, pp. 38–40, 1982.

9. _____, Technical Writing for the Computer Sciences, *Technical Writing Teacher*, 10:3, pp. 85–91, 1983.

10. V. M. Arms, Engineers Becoming Writers: Computers and Creativity in Technical Writing Classes, in *Writing at Century's End: Essays on Computer-Assisted Composition*, L. Gerrard (ed.), Random House, New York, pp. 64–78, 1987.

11. E. Kelly, Processing Words and Writing Instructions, in *Writing at Century's End: Essays on Computer-Assisted Composition*, L. Gerrard (ed.), Random House, New York, pp. 27–33, 1987.

12. N. Allen, D. Atkinson, M. Morgan, T. Moore, and C. Snow, What Experienced Collaborators Say about Collaborative Writing, *Journal of Business and Technical Communication*, 1, pp. 70–90, 1987.

13. M. Morgan, N. Allen, T. Moore, D. Atkinson, and C. Snow, Collaborative Writing in the Classroom, *Bulletin for the Association of Business Communication*, 50:3, pp. 20–26, 1987.

14. C. Ruenzel and J. E. Sheldon, Education and the New Technical Writer, unpublished paper, University of Wisconsin–Milwaukee, May 15, 1987.

15. J. Tanner, Manager of Customer Documentation, M & I Data Services, Personal Interview, Brown Deer, Wisconsin, April 22, 1988.

16. C. Ruenzel, Group Leader of Systems Documentation, Deluxe Data Systems, Personal Interview, Brown Deer, Wisconsin, April 22, 1988.

17. H. S. Wiener, Collaborative Learning in the Classroom: A Guide to Evaluation, *College English*, 48:1, pp. 52–61, 1986.

18. K. Bruffee, Liberal Education and the Social Justification of Belief, *Liberal Education*, 68, pp. 95–114, 1982.

19. F. M. O'Hara, Jr., Beyond Word Processing: Computers in the Composition Process, *Word Processing for Technical Writers*, R. Krull (ed.), Baywood Publishing Company, Inc., Amityville, New York, pp. 72–85, 1988.

20. T. Weiss, Word Processing in the Business and Technical Writing Classroom, *Computers and Composition*, 5:2, pp. 57–70, 1988.

21. D. P. Balestri, Softcopy and Hard: Word Processing and Writing Process, *Academic Computing*, 2:5, pp. 14–45, 1988.

22. W. Perry, *Forms of Intellectual and Ethical Development in the College Years: A Scheme*, Hold, New York, 1970.

23. C. Gilligan, *In a Different Voice: Psychological Theory and Women's Development,* Harvard University Press, Cambridge, 1982.

24. M. F. Beleky, B. M. Clinchy, N. R. Goldberger, and J. M. Tarule (eds.), *Women's Ways of Knowing: The Development of Self, Voice, and Mind,* Basic Books, New York, 1986.

25. P. Bizzell, William Perry and Liberal Education, *College English, 46*:5, pp. 447-454, 1984.

26. J. R. Goldstein and E. Malone, Journals on Interpersonal and Group Communication: Facilitating Technical Project Groups, *Journal of Technical Writing and Communication, 14*:2, pp. 113-131, 1984.

27. K. Bruffee, *A Short Course in Writing,* 3rd Edition, Little Brown, Boston, 1985.

28. K. Spear, *Sharing Writing: Peer Response Groups in English Classes,* Heinemann, Portsmouth, New Hampshire, 1988.

29. M. A. Janda, Talk Into Writing: How Collaborative Writing Works, Dissertation, University of Illinois at Chicago, Chicago, 1988.

30. L. Vygotsky, *Thought and Language,* MIT Press, Cambridge, 1978.

31. K. Bruffee, Collaborative Learning and the "Conversation of Mankind," *College English, 46,* pp. 635-652, 1984.

32. E. Tebeaux, Redesigning Professional Writing Courses to Meet the Communication Needs of Writers in Business and Industry, *College Composition and Communication, 36*:4, pp. 419-428, 1985.

33. J. Selzer, The Composing Process of an Engineer, *College Composition and Communication, 34*:2, 1983.

34. J. Trimbur, Collaborative Learning and Teaching Writing, in *Perspectives on Research and Scholarship in Composition,* B. McClelland and T. Donovan (eds.), Modern Language Association, New York, pp. 87-109, 1985.

35. G. Broadhead and R. Freed, *The Variables of Composition: Process and Product in a Business Setting,* Southern Illinois University Press, Carbondale, p. 131, 1986.

PART IV:

Current Industrial Concerns: Gathering, Verifying, and Editing Information

Successful technical communicators must be experts in gathering information, checking on the accuracy of that information, and soliciting and responding to the suggestions of others during the production of that information. These collaborative activities ensure quality documents. In turn, these documents, particularly within the laboratory or the computer field, accompany or have become the actual "products" of industry. The three authors in this section use their industrial experiences to reflect upon how technical communicators can best engage in these collaborative activities. Generally, their descriptions and recommendations are based on a central mission: to ensure that writers, editors, graphic designers, and technical experts work together early and often during the production of a document and maintain a cooperative rather than competitive spirit. In fact, if all three authors were writing a job ad for a technical communicator, the first skill listed would be the ability to collaborate.

The three chapters offer a new and practical look at industrial collaborative activities. Since technical communicators must interview and gather complete information from technical experts, Hickman offers a system not only for discovering when and why that information might be incomplete but also for establishing a rapport with technical experts. Grice defines what characteristics of information should be verified—checked for accuracy—and who should check and when they should do so. Shirk weighs peer and hierarchical editing in both industry and academia to select when each should be used and to suggest a new style of editing—collaborative editing. While it would be tempting to say these three essays represent a progression of collaborative activities—one first gathers, then verifies, and finally edits information, again the authors reveal that collaboration is on-going and frequent, recursive and not linear. All three authors investigate how to "manage" the industrial collaborative process—whom to involve, when to involve them, and how to make sure that their involvement is fruitful.

Dixie Hickman, in her "Neuro-Linguistic Programming Tools for Collaborative Writers," teaches technical communicators how to "read" certain language patterns and non-verbal signals of an interviewee. By learning the Meta-Model, technical communicators can discern distortions or incompleteness in the information an interviewee offers. Representational systems and sensory data bases provide clues as to what strategies technical experts used to complete the tasks they describe. These systems also help the interviewer "see" the technical expert's world picture so the rapport between interviewer and interviewee is strengthened, or, as Kenneth Burke would say, "consubstantiation" is established. Hickman's suggestions are particularly useful when interviewing technical experts who are so experienced that they have forgotten what the novice user needs to know. "Learning to use these NLP tools systematically in our analysis of utterance can make even the best 'natural' interviewer better," Hickman claims. She uses sample dialogues between technical experts and writers to support her suggestions.

In "Verifying Technical Information: Issues in Information-Development Collaboration," Roger Grice links collaborative activities with an essential industrial task—making sure that information that accompanies a product is accurate and useful. The technical communicator or "information developer" collaborates with editors and technical experts to verify four aspects of information: accuracy and completeness, usability, suitability for audience, and presentation and style. Grice indicates when and by whom these aspects of information are checked during four collaborative activities: editing, reviewing/ inspecting, usability testing, and ensuring early customer involvement. The "inspection meeting," in which collaborators debate, negotiate, and settle the content and style of information, takes the place of traditional review or editing processes. Also, through "early support systems," selected customers get the product and information before the general market does, and these customers collaborate with corporate technical writers and editors to verify the accuracy of the information. Verification activities, during which people both inside and outside the organization collaborate, then are "integral parts of information development."

While Grice describes the editing process as an aspect of verification, Henrietta Shirk in "Collaborative Editing: A Combination of Peer and Hierarchical Editing Techniques" reports the results of an in-depth study of editing in two settings: academia and industry. Students and technical communicators in these environments report their impressions of peer editing, or editing performed by a writer's equal, and hierarchical editing, or editing initiated by someone higher than the writer in the organizational structure. Shirk describes the strengths and weaknesses of both types of editing, and rationales for and the environments in which each is used, and the skills that ensure the success of both. While most students had experienced hierarchical editing throughout their educational life, they found that peer editing, conducted in a supportive setting, gave them insight on audience reaction and began to help them overcome resistance to giving and receiving criticism about their writing. Within industry, Shirk found that "knowing how to orchestrate the two collaborative processes [of peer and hierarchical editing] can become the measure of a manager's and an organization's success." And this "collaborative editing" or combination of both activities requires diplomacy and interpersonal skills on the part of both writer and editor.

CHAPTER 10

Neuro-Linguistic Programming Tools for Collaborative Writers

DIXIE ELISE HICKMAN

Collaboration between a technical expert and a technical writer who may or may not know anything about the subject matter is common in industry. In this kind of collaboration, the writer is usually responsible for most of the written document, but the two cooperate in generating the content. The technical writer who is familiar with two sets of tools from Neuro-Linguistic Programming (NLP) can more smoothly and efficiently gather and interpret information while interviewing the technical expert: the Meta-Model for correcting gaps or distortions in the information and representational systems for understanding the expert's model of the world. Familiarity with these tools allows an interviewer to be systematic in analyzing the utterances of the interviewee, and practice with the tools soon renders their use automatic.

NEURO-LINGUISTIC PROGRAMMING

Neuro-Linguistic Programming, developed by John Grinder and Richard Bandler, is a model of human communication that analyzes the structure of subjective experience [1-4]. Its sources—and its criteria—are whatever works (with testable results). Consequently, it has roots in and applications to a great many aspects of human communication.

A major aspect of human communication is language, which is both systematic and idiosyncratic. Our language reflects the way we perceive our world.

Becoming aware of certain patterns and observing how they are used in other people's language can give us a window on their world, a grasp of their reality.

THE META-MODEL

The Meta-Model identifies certain relationships between the surface structure of language and the "deep structure" [1], where lies the fundamental meaning behind the utterances we hear. Our observations of violations of this Meta-Model help us isolate certain predictable patterns of language that reflect some distortion or incompleteness of information. For example, when a technical expert exclaims, "Nobody can understand what I'm doing!" the collaborating writer can either give up the project or recognize that the expert's language indicates some distortion (surely *somebody*, even if it is only the expert, is capable of understanding) or some incompleteness (perhaps the expert means that nobody without certain background information can understand).

Since it is, for all practical purposes, impossible to tell everything about an experience, some selectivity is inevitable. Therefore, in most conversational situations, Meta-Model violations are natural and inconsequential. In a technical situation, however, when a writer is collaborating with an expert to produce a technical document, such violations can be significant. For instance, important elements of a procedure may have become so automatic for the expert that they are no longer conscious steps. Also, long experience may have produced refinements in judgment that the expert may never have consciously articulated. The writer needs access to these details which the expert may automatically gloss over, for the audience of the document to be produced will not have the expert's experience. While interviewing the expert, the writer's ability to identify and interpret these patterns provides clues to potential trouble spots and prompts questions to be asked or answered: the answers to these questions may provide a fuller or more objective picture of the reality. The Meta-Model is described below, and excerpts from a dialog in a collaborative writing situation illustrate the patterns.

Bandler and Grinder's Meta-Model may be divided into three major categories: Gathering Information, Expansion of an Individual's Model of the World, and Semantic Well-Formedness [3]. Each of these categories is further subdivided. When a writer observes a violation in any of these categories, the patterns indicate a deletion or restriction of some of the information, a limitation of the speaker's model of reality, or a distortion of meaning.

Patterns Pointing to Incompleteness in Gathering Information

A writer interviewing an expert naturally is concerned with gathering all the pertinent information. Three Meta-Model violations point to some omission in the expert's utterance: deletion, unspecified verbs, and nominalizations. Of course, many violations may be insignificant, but the writer who is alert to the possibility of the omission of useful information will do less backtracking later.

Deletions — Often the writer can guess what information has been omitted from the expert's statements, but if there is any possibility of ambiguity or uncertainty, the writer can elicit the missing information by asking "what, specifically."

In a *Simple Deletion*, some object or person is omitted from the surface structure of the utterance.

> Ex. Expert (E): Once I get a picture of the problem, I can design an algorithm. Getting that first picture is just a matter of focusing.
> Writer (W): Focusing on what, specifically?

The Writer recognizes that the picture must be stimulated by something and asks for the missing prompt to the design.

In a *Comparative Deletion*, the comparative referent is deleted.

> Ex. E: Using a *while loop* is a better way to code.
> W: Better than what other ways?

The Writer asks for the comparison to be made specific. The Expert's utterance also contains a simple deletion which may contain a key element in deciding a programming design. Once the comparison is specified, the Writer will probably continue:

> W: The *while loop* is better for coding what, specifically?

In *Lack of Referential Index*, the object or person being referred to is unspecified.

> Ex. E: I've done some studying on this. They say you shouldn't use GOTO.
> W: Done some studying on what, specifically?

The Writer checks whether *this* refers to the GOTO pattern only, to a broader aspect of program design, or to the pattern in this particular context or kind of context.

> W: Who, specifically, says you shouldn't use GOTOs?

Challenging the second lack of referential index in the utterance elicits the authority for the statement.

Unspecified verbs — A verb may be too general, or too abstract, to describe the specific, concrete action needed. Asking "how, specifically," or "in what way" about the verb will elicit the missing information.

> Ex. E: Having the timer running helps me know where to start.
> W: How, specifically, does the timer help?

Help is one of those general words that may cover many kinds of actions, but even more slippery are the abstract verbs, like *know*.

Ex. E: I always know which part to do first.
W: How, specifically, do you know?

People "know" in many ways. This Expert may "know" that the picture of the problem is complete or that established procedures dictate the starting point.

Nominalizations – In a nominalization, an ongoing process is represented as static. When a process is represented as a completed event, access to the steps in the procedure is limited. In a collaborative writing situation, a nominalization often signals a subprocedure which may need explanation.

Three tests can identify a nominalization. First check all non-verb words in the utterance. If you can think of a verb that is similar in sound or form and in meaning, you probably have a nominalization. Using the first sentence of this paragraph as an example, we would think of the verbs *test* and *nominalize.* Second, if no similar verb comes to mind, try the word in this phrase: "an ongoing _____." If the phrase makes sense, a process has been nominalized. We can easily imagine ongoing processes of testing and nominalizing. If you need a third test, imagine the thing represented by the word in a wheelbarrow; if you can't put it in a wheelbarrow, it is probably a nominalization. We can imagine certain kinds of tests riding in a wheelbarrow, but trying to picture a nominalization sitting in the wheelbarrow will probably draw a blank. To retrieve the process behind the nominalization, rephrase the nominalization using a verb and ask "about what," or "how, specifically."

Ex. E: The initialization takes a few minutes.
W: What get initialized? How do you initialize whatever it is?

Initialization doesn't fit into the wheelbarrow, and the "ongoing initialization" takes a few minutes, so the Writer uses the verb form *initialize* to uncover the process.

Patterns Pointing to Limitations of a Person's Model

Especially useful in problem solving, challenging violations of this segment of the Meta-Model can reveal unnecessary restrictions on the ways people think about their experience. Four patterns invite challenge: presuppositions, modal operators of possibility and necessity, complex equivalents, and universal quantifiers.

Presuppositions – Implicit assumptions in people's models of the world show up in their language. Left unchallenged, these presuppositions may limit choices of action unnecessarily. A general test for presuppositions is to ask, "On what, if anything, unstated, does the truth of this utterance hang?" Certain patterns of language involve presuppositions, ranging from simple presuppositions about existence ("I used the VAX" presupposes that a VAX computer exists) and availability ("Use the VAX" presupposes that the VAX is available for use) to more complex presuppositions ("If this third version of the program works, I'll be surprised" presupposes that a program exists, that it has been tried two times already, and that it has not worked). For a listing of some of these environments, with examples, see Bandler and Grinder's *The Structure of Magic* (1, Appendix B, Vol. I). Many presuppositions are, of course, quite sensible, but some may indicate unnecessary limitations.

> Ex. E: If all of your questions are going to need such detailed thinking, we need to stop for a moment so I can start the next phase of this process.

In the general test for a presupposition, for it to be true that the dialog needs to stop, the questions must require detailed thinking. The statement also presupposes that the process the Expert is engaged in has another phase, that the Expert must initiate it, and that the dialog would interfere with the action.

> W: I do seem to be asking for a lot of detail, don't I? Go ahead and do what you need to do. I'll look back over the notes I've made so far to see if I need to ask more questions. If I get all the details now, I'll be able to present your work correctly without having to come back and interrupt you later.

The Writer accepts as a sensible presupposition that the Expert must start a new part of some process without distraction and addresses the Expert's underlying concern about the length of the interview by confirming the presupposition about the detail being asked for.

Modal operators of possibility or necessity — Certain verb forms (e.g., can't, must, should) point to "rules" about behavior. In problem solving, these words may signal unnecessary restrictions which the writer may want to correct. In procedures, they may indicate points where the novice reader will need cautions or warning flags. The Writer can challenge these Meta-Model violations by asking "what will happen if you do/don't."

> Ex. E: You should write the documentation as soon as you finish the code. You can't turn that over to somebody else.
> W: What will happen if you don't write the documentation? What will happen if you wait to write it?

The Writer has two points to check the modal operator against. First, is it necessary to write the documentation, and then must it be done immediately on finishing the code. In the Expert's second sentence, *that* lacks a referential index, and the sentence also contains a negated modal operator of possibility.

> W: What prevents you from turning the documentation over to somebody else? What would happen if you did?

Rather than simply assuming that *that* refers to documentation, the Writer verifies it by repeating the phrase and supplying the reference and then directly challenges the negative modal operator *can't*.

Complex equivalent – In complex equivalent, two experiences or two uses of the same verb are considered synonymous, an equation which may not be warranted. The two experiences are often expressed in back-to-back sentences with the same grammatical structure. The Writer can verify the equation by asking if X (sentence one) always means Y (sentence two).

> Ex. E: She's carrying her clipboard. She's going to find something to complain about.
> W: Does her carrying the clipboard always mean she'll complain?

This application of complex equivalent is especially useful in situations involving interpersonal relations and procedures dependent on specific people. An application more common in technical writing recognizes the need to be cautious in understanding the Expert's complex equivalent for commonplace verbs.

> Ex. E: I hate to write in Pascal.

The Writer may want to discover the Expert's complex equivalent for *hate*. Is it the same as hating to write with a pencil or hating to eat anchovies on a pizza? Do Expert and Writer mean the same thing when they say *hate*? Is the Expert's hating to write in Pascal as strong as the Writer's hating to muck out the barn while everyone else goes shopping? Perhaps the Expert "hates" Pascal because it seems inadequate for a particular programming task. Or perhaps the Expert is seldom required to use Pascal and so is simply uncomfortable with an unfamiliar language.

Universal quantifiers – Certain words (e.g., always, never, every) generalize limited experience to a broader class of experience. While often valid generalizations, they may also obscure important exceptions. Let's return to a previous example.

> Ex. E: I always know which part to do first.
> W: Is there never a time when you don't know what to do first?

In the previous situation, the Writer zeroed in on the unspecified verb. This time, the Writer asks for exceptions to the universal quantifier. The question may uncover more of the Expert's strategy for "knowing," the kind of details the Expert interprets automatically but which could be essential for a novice to consider consciously to make a correct judgment. The question may also uncover special circumstances in the procedure which may need to be mentioned in a troubleshooting guide.

Patterns Pointing to Semantic Ill-Formedness

Violations of this section of the Meta-Model involve three types of illogical semantics: cause-effect, mind-reading, and lost performatives.

Cause-effect — In a cause-effect violation, a causal relationship is assumed between two events which may not be connected—the old *"post hoc, ergo propter hoc"* logical fallacy.

> Ex. E: If I don't code the output routine first, the program always gets messed up, and I have to start over.
> W: How does not coding the output routine first mess up the program?

In this example, the Writer challenges the cause-effect implications to retrieve the positive reasons for coding the output routine first. Absence of the output routine code does not in itself mess up the program, but having the routine done first makes the rest of the coding easier. The Writer may also want to ask "how, specifically" about the unspecified verb "mess up" and to challenge the universal quantifier with, "Does not having the output routine code ALWAYS mess up the program? Has there ever been a time when it didn't mess up the program?"

Mind-reading — In the mind-reading violation, a person presumes to know without being told what another person thinks. The challenger seeks the person's experiential basis for the assumption.

> Ex. E: You really enjoy asking these nitpicky questions, don't you! And you already understand what I mean, anyway. Well, users like inverted video windows when they have to key in data.
> W: Which users, specifically? And how do you know they like it?

The Writer chooses to ignore the mind-reading in the first two sentences. (The Writer's smile may indicate pleasure in the questioning but may also be intended to encourage the Expert to reply pleasantly; the Writer's understanding may or may not match the Expert's understanding, and more specific data is needed to verify a match.) Instead, the Writer asks for a referential index for "user" and for the Expert's experiential basis for assuming that the user will like inverted video windows. For example, has the Expert conducted tests or gathered client testimony?

Lost performative — Here, individual experience is generalized into a universal value judgment. The speaker presents a personal opinion as if it were generally accepted. What is "lost" is who made the judgment and what criteria were used.

> Ex. E: Programmers shouldn't write user's manuals.

The Writer has only the Expert as the source of this pronouncement. In the underlying "deep structure" of the utterance, the Expert says, "I say to you that programmers shouldn't write user's manuals." The Writer will question to find the source of this belief. Is this a simple idiosyncratic distaste for writing manuals, a generalization based on one or a few bad experiences, or a conclusion drawn from careful consideration of unfavorable consequences? In short, challenging the lost performative helps the writer distinguish between Expert belief and Expert knowledge.

REPRESENTATIONAL SYSTEMS

A second valuable tool that writers can use in a collaborative situation is representational systems [1, Vol. II]. Information is gathered into our brains via our senses. Retrieval of that information has some connection still with those senses. One language clue to the way a person processes information is the sensory system represented in the process words, or predicates (verbs, adjectives, and adverbs). For a collaborative writer, the representational systems have two main uses. One is to help the Writer achieve rapport with the Expert [4]. A second is to facilitate "unpacking" or translating the Expert's strategy for accomplishing a specific task [2].

Identifying Representational Systems

Information enters the brain through our primary senses. Storage and retrieval of this information retain some connection with the input mode. Certain tasks put demands on specific sensory channels. For example, touch typing is a kinesthetic activity, while sorting computer chips is primarily visual. Other tasks, like remembering where you left your keys, can be processed in any of several channels, and the choice is a personal preference, reflective of the individual way you structure and represent your experience to yourself.

People reveal the sensory channels they use to process data by their language [1, Vol. II] and by their eye movements [3-4]. In language, the process words, or predicates (verbs, adjectives, and adverbs), often mirror the mental process. For instance, when someone says to you, "To see results, you'll have to focus on each part and then picture it from a broad perspective," you have two pieces of information about that person's model of how one gets results. First, you know that the person examines each part or step individually and then combines them into a whole. Second, you know the process this person uses to get results is primarily visual, since the utterance is full of visual predicates: see, focus, picture, perspective.

Of the six possibilities, people's predicates most often fall into one of four categories. Visual (words like *see, picture, look, focus, inspect,* and *observe*), auditory (words like *hear, listen, say, tell, speak, click,* and *sound*), kinesthetic (words like *feel, touch, grasp, handle, run, push,* and *catch*) are the most common, followed closely by unspecified predicates, words that do not indicate any specific sensory system (such as *seem, aware, investigate, know, understand, contemplate, believe,* and *think*). Less frequently, we hear gustatory predicates (words like *taste, sour, sweet,* and *bitter*) and olfactory predicates (words like *smell, scent, nose,* and *fragrant*). Though rare, when they occur, gustatory and olfactory predicates may be very important.

Eye movements are a second indicator of the mental process someone is using [3]. When you access information stored in your brain, your eyes reflect the sensory channel(s) you use to retrieve that information. If you are a typical right-handed person, when you recall something visual, your eyes probably move upward and to the left. When constructing a visual image, your eyes move up and to the right. Sometimes, when you are concentrating on something visual in your mind, you defocus your eyes and stare into space directly in front of you, some distance away. Similarly, when you remember something auditory (how does that song go?), your eyes move directly to the left. When imagining how something new might sound, your eyes move directly to the right. When you are deeply in touch with your feelings (or some taste or smell), your eyes will mirror your kinesthetic involvement by moving down and to your right. When you are having an internal dialog (talking to yourself), your eyes will be down left. Left-handed people are often wired neurologically in the reverse, though not always. And occasionally, one channel will reverse positions for constructed and recalled information (see Figure 1 for a map of basic eye positions for a right-handed person).

Becoming aware of these two ways (eye movements and sensory predicates) people demonstrate their internal processes helps us understand the ways they represent their experience, the ways they comprehend and interpret the world. For persons in a collaborative writing situation, this awareness serves two specific purposes: it helps us foster rapport, and it helps us unpack internal strategies for certain tasks.

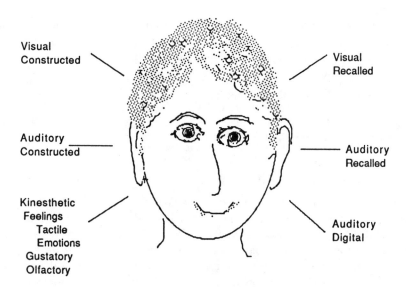

Figure 1. Eye movement accessing cues.

Matching Representational Systems for Rapport

When two people have different methods of making sense of the world, they often find it difficult to work together. Some of these differences are apparent in the predicates each person uses. Someone trying to *tune in* to what we're saying will have difficulty if we keep trying to *paint the big picture*. When we match our partner's model of the world, we improve rapport and communicate more effectively.

Consider the situation in the following dialog:

> W: How do you know when the reaction is complete and the beaker should be immersed?
> E: Hmm. [Eyes move up left and then down right.] I'm not sure I can put a finger on it. You just get a feel for the timing.
> W: That sounds like you kind of tune in to the reaction itself.
> E: Well, no, it's a bigger handful than that. You have to grasp the whole process and the way the parts fall together.
> W: Then would you say it's, maybe, a set of clicks as one thing calls for another?
> E: No. You know, maybe we'd better drop this project. You're not really in touch with this kind of chemistry.

The problem here has nothing to do with the technicalities of chemistry; it's the chemistry between the two collaborators. They think primarily in different representation systems (W, auditory; E, kinesthetic), and neither can *see* things from the other's viewpoint.

If either were aware of the language manifestation of the problem, that one could switch representation systems to match the other, thus establishing the rapport necessary to deal with the technical details.

> W: Wait. Let's run through it again, and let me try again to get hold of it.
> E: OK. Things bubble along, sort of bouncing on the burner, and when the tension reaches a certain level, you grab it.
> W: I think I'm getting a handle on it. The process has its own rhythm, and you kind of move with it, right?
> E: Hey, you're catching on! Now the next part . . . .

Unpacking the Expert's Strategy

The Expert's explanation of how a person knows when a particular phase of a process has run its course (you just get a feel for it) illustrates the lack of conscious detail in most experts' strategies. While much of an expert's strategy may be unconscious, developed by long experience, the document the Writer is trying to produce is being written for a novice who wants to learn the strategy by deliberate steps. In addition to using the Meta-Model to get more specifics about the verb, the Writer may want to explore whether feeling is a necessary part of the process or whether the judgment might as effectively be based on "It just looks right" or "Something clicks."

Let's *rewind* the previous dialog and play it with a different *tune*, changing the *scene* a little:

E: Hmm. [Eyes move up left and then down right.] I'm not sure I can put a finger on it. You just get a feel for the timing.

W: How do you get that feel? What feeling, specifically?

E: [Eyes down right.] Well, it's almost a compulsion. I just sense it's time to grab it.

W: How do you know when to get the feeling? What happens right before you get it?

E: [Eyes move up left and then down right.] Nothing, really. I'm just standing there, waiting for the rhythm to peak.

W: OK. You know within about two minutes how long it's going to take. So you're standing there already. Are you watching the mixture cook? [E's eye movements suggest a visual prompt to the feeling.]

E: Yeah, I'm watching the bubbles and sorta grooving on the rhythm of the bubbles.

W: And this rhythm peaks? How, specifically, do you know when it's reached the peak?

E: Well, you can see them moving faster. And then they get smaller and faster, tighter and more intense.

W: So you watch the rhythm of the bubbles when it's about time for the reaction to peak. When they get faster, you know it's almost time. And when you see them get smaller and faster, you know you have to grab it quick.

E: You may have hit the nail on the head. Let's check it out.

This time the Writer repeats the kinesthetic predicates, rather than switching representation systems, while using the Expert's eye movement cues to suggest possible elements in the strategy, and uses the Meta-Model to elicit more specific information, more verifiable sensory data. The result (which will, of course, have to be tested to make sure other clues are not being overlooked) is a clear description of the strategy the Expert uses. The visual landmarks serve as guides until the novice develops expertise and acquires "a feel" for the process.

CONCLUSION

In more recent publications concerning Neuro-Linguistic Programming, the linguistics of the Meta-Model seems less popular than the representational systems, perhaps because recognizing the sensory channels indicated by predicates and watching eye movements require less study or less sensitivity to language than mastering the Meta-Model patterns. Nevertheless, the Meta-Model will always remain a primary analytical tool, the bedrock on which many of the later techniques are founded.

Neuro-Linguistic Programming researchers have made the structure and system of interactive interpersonal communication more explicit. Good inter-viewers use many of these techniques naturally. Learning to use these NLP tools systematically in our analysis of utterance can make even the best "natural" interviewer better. Indeed, communication sometimes improves so dramatically as to make NLP seem almost "magic." These tools not only help us cover all the bases, but they also suggest alternative routes to the information we seek. Only a few of the many techniques available could be demonstrated here. The books listed in References and Selected Bibliography are valuable resources for communicators.

SUGGESTED READING

The Reference section of this article cites books dealing with the foundation and fundamentals of NLP. The Selected Bibliography lists additional books about NLP. The list is by no means exhaustive; it reflects my own preferences and prejudices and my opinion of what would be most useful. Here are a few recommendations to guide your initial selections:

- For a comfortable introduction, try *Frogs into Princes, Meta-Cation I,* or *Practical Magic.*
- For emphasis on language, see *Structure of Magic I* and *II.*
- For the technical research behind NLP, see Robert Dilts' books.
- For applications to teaching, see Jacobson and Cleveland first.

REFERENCES

1. R. Bandler and J. Grinder, *The Structure of Magic,* Vols. I and II, Science and Behavior Books, Palo Alto, California, 1975 and 1976.
2. R. Dilts, J. Grinder, R. Bandler, J. DeLozier, and L. Cameron-Bandler, *Neuro-Linguistic Programming I,* Meta Publications, Cupertino, California, 1979.
3. R. Dilts, *Roots of Neuro-Linguistic Programming,* Meta Publications, Cupertino, California, 1983.
4. _____, *Applications of NLP,* Meta Publications, Cupertino, California, 1983.

SELECTED BIBLIOGRAPHY

General NLP Reading

Bandler, R. and J. Grinder, *Frogs into Princes,* Real People Press, Box F, Moab, Utah, 1979.

Bandler, R., *Using Your Brain—for a Change,* C. Andreas and S. Andreas (eds.), Real People Press, Box F, Moab, Utah, 1985.

Grinder, J. and R. Bandler, *Reframing: Neuro-Linguistic Programming and the Transformation of Meaning,* Real People Press, Box F, Moab, Utah, 1982.

Gordon, D., *Therapeutic Metaphors: Helping Others Through the Looking Glass,* Meta Publications, Cupertino, California, 1978.

Lankton, S. R., *Practical Magic: The Clinical Applications of Neuro-Linguistic Programming,* Meta Publications, Cupertino, California, 1979.

Lewis, B. A. and R. F. Pucelik, *Magic Demystified: A Pragmatic Guide to Communication and Change*, Metamorphous Press, Lake Oswego, Oregon, 1982.

Yeager, J., *Thinking about Thinking with NLP*, Meta Publications, Cupertino, California, 1985.

Special Applications of NLP

Cameron-Bandler, L., *Solutions: Practical and Effective Antidotes for Sexual and Relationship Problems*, Future Pace, Inc., San Rafael, California, 1985.

Cleveland, B., *Master Teaching Techniques*, The Connecting Link Press, Stone Mountain, Georgia, 1984.

Grinder, J., and R. Bandler, *Trance-Formations: Neuro-Linguistic Programming and the Structure of Hypnosis*, Real People Press, Box F, Moab, Utah, 1981.

Harper, L., *Classroom Magic: Effective Teaching Made Easy*, Twiggs Communication, Troy, Michigan, 1982.

Jacobson, S., *Meta-Cation*, Vols. I, II, and III, Meta Publications, Cupertino, California, 1983, 1986, and 1988 ("Prescriptions for Some Ailing Educational Processes").

McMaster, M. and J. Grinder, *Precision: A New Approach to Communication*, Precision Models, Beverly Hills, California, 1980. (Business applications.)

Van Nagel, C., E. Reese, R. Siudzinski, and M. Reese, *Megateaching and Learning—Neuro-Linguistic Programming Applied to Education*, Southern Institute Press, Inc., Indian Rock Beach, Florida, 1985.

CHAPTER 11

Verifying Technical Information: Issues in Information-Development Collaboration

ROGER A. GRICE

Verification is one of the most important steps in producing good technical information; it rightly makes up a major portion of the information-development cycle. It is also an activity that cannot be done alone; just as people have trouble proofreading their own writing, so too are people who produce technical information not in the best position to evaluate that work thoroughly and objectively.

There is an increased interest in and awareness of verifying technical information. Why? A major reason is that technology and "high tech" products are becoming a part of everyday life. This diffusion of technology means that there is an increasingly wide use of technical information by those who are not technically skilled in the use of the products or the technology described by the information. Also with the ever-increasing demands being placed on people's time, users need to read and apply technical information efficiently and effectively. Both of these factors demand that technical information be verified as completely and thoroughly as possible before it is presented to users.

The types of verification activities described here are integral parts of an information-development process that, in turn, is an integral part of an overall product-development process. The steps of the information-development process, in fact, parallel those of the processes followed by the professionals—engineers and programmers, for example—with whom information developers work. The verification activities thus involve close collaboration between information developers and other product developers, all of whom are following similar processes on similar schedules to achieve a common goal—delivery of a useful and usable product. Information development, and consequently, information verification could, of course, be done as an activity completely separate from product development (for example, a programming manual or product user's guide developed by someone outside of the product-development group), but this type of information development and verification is not discussed here.

What is involved in verifying technical information? To do a thorough job requires the coordinated efforts of many people, each using specific skills and examining the information from unique viewpoints. This chapter examines major verification activities, which emphasize the need for coordinated collaboration among those concerned with the information and the product that the information describes. The section "Verifiable Characteristics of Information" discusses the characteristics of information that can be verified and when, during the information's development, these characteristics can be verified. Then "Major Classes of Verification Activities" describes four types of verification in depth, focusing on how writers collaborate to ensure the quality of the verification.

VERIFIABLE CHARACTERISTICS OF INFORMATION

Different types of verification activities lend themselves to verifying different characteristics of information. Verification of the following information characteristics is essential:

- accuracy and completeness of the information;
- usability of the information to perform a specified task;
- suitability of the information for its intended audience; and
- presentation and style and their relationship to quality and usability.

A thorough information-verification program examines each of these characteristics from several points of view. And while certain activities may give closer scrutiny to some characteristics than others, all four classes of characteristics should be considered during any verification activity. For example, while a technical review and inspection may focus primarily on accuracy and completeness, it should not ignore the usability, suitability, presentation, or style of the information.

Verifying each of these information characteristics requires writers to work with other people, or groups of people, to ensure that the verification is done properly and effectively (see Table 1).

Table 1. Collaborative Relationships for
Verifying Technical Information

Information Characteristic	Person or Group with Whom Writer Collaborates	Verification Activities
Accuracy and completeness	Technical experts	Review and inspect
	Other writers	Ensure complete, consistent coverage
	Editor	Edit to ensure consistent coverage and treatment
	Test personnel	Test to ensure accuracy and completeness
Usability for a task	Test personnel	Test for usability
	Editor	Edit for usability
	Users or user surrogates	Ensure understanding of user requirements; test for usability
Suitability for an audience	Users or user surrogates	Review and discuss the information
	Editor	Edit from audience's perspective
	Marketing representatives	Gather user requirements; make plans to meet those requirements
Presentation and style	Editor	Edit for literary quality and adherence to standards

MAJOR CLASSES OF VERIFICATION ACTIVITIES

Wagner cites twelve types of verification activities that he used and observed as an IBM information developer: editing, lead-writer editing, peer review, self-review, technical-owner review, reviews by people with special skills, testing, walkthroughs, design reviews, checklists, mechanical reviews, and user feedback [1]. In this chapter, I group these activities into four categories:

1. **Editing,** in which an editor scrutinizes the draft for style, structure, and adherence to standards and conventions.
2. **Reviewing/Inspecting,** in which groups of product developers (technical experts) and product support people (such as those who provide

maintenance and service to customers) review the content of the information presented and the manner in which it is presented.

3. **Usability Testing,** in which groups of users or user surrogates use early versions of the product, including the product information, in a controlled setting to determine its accuracy, suitability, and usability.

4. **Ensuring Early Customer Involvement,** in which selected customers use the product and its associated information in their own environment and in the way they would use the final product.

WHEN VERIFICATION IS DONE/ WHAT CAN BE VERIFIED

Verification is often thought to be something that is done on a completed product or piece of information. This is not the case. Information verification and testing can, and should, be done at each stage of its development, from prototype testing, to iterative testing during development, to after-the-fact verification through surveys. The focus of the verification and the attributes that can be verified vary through each phase of the information-development cycle.

The Role of Editing in the Information-Development Cycle

Editing can be done on any level of draft. In fact, great benefits may be derived from having an editor initially evaluate a book's outline to prevent the introduction of organizational or stylistic flaws that would need extensive correction if they were first uncovered later in the information-development cycle.

Editing of a first draft generally focuses on the writer's approach to the subject matter and the audience. Since much of the material present in the first draft may be substantially changed, or even removed, by the time the final draft is produced, the editor can be most effective by examining trends in the writing and discussing improvement methods with the writer rather than editing the draft word-for-word.

Editing of a final draft must focus directly on the presentation of material. The editor must examine the draft carefully for approach to the material, relationship of the draft to drafts of other information units that deal with the same subject or related subjects, and for style, word choice, and grammar.

At each stage that a draft is edited, writer and editor must work together to ensure that editorial changes are understood by the writer, that the editing changes do not change the technical meaning of the information nor introduce technical inaccuracies, and that the writer has answered all questions raised by the editor.

The Role of Reviewing/Inspecting
in the Information-Development Cycle

The inspection process, as described by Fagan for programming processes can be used to great advantage within an information-development process [2]. The inspection process, which builds on methodology used for inspecting the quality of manufactured objects, shifts the emphasis in reviewing from distribution of drafts to reviewers for their comments (which are often mailed back to the writer) to an inspection *meeting* for which participants must prepare ahead of time (by reviewing the draft). The meeting is run under the leadership and direction of a moderator. In the inspection meeting they participate actively by discussing the comments and concerns uncovered in their review and resolving differences of opinion among participants. After the meeting they perform required follow-up work as agreed to during the inspection meeting.

The publications inspection procedure (summarized in Table 2) is an effective way to ensure the adequacy of reviews. Each step of the procedure is divided into phases to ensure that a proper and thorough job is done at each step.

Some of the collaborative relationships presented in Table 2 are of a long-lasting nature; a writer may work with one editor over the course of developing a manual or other unit of information and even continue the collaboration through a series of revisions and updates. Other relationships, however, are transient. A product developer may be involved with a single inspection of a piece of information and then have no further involvement at all.

Information can—and should—be reviewed more than once during its development. To ensure the adequacy of these reviews, a group of related inspections is defined, specifying the timing and purpose of the inspections to be held to ensure proper and thorough review. A set of inspections for reviewing information during its development are defined below. Of course, if information is being developed on a different schedule or by using a different process, the set would be modified accordingly.

- The Inspection Planning Meeting: At this meeting, all those who will be involved in the information inspections gather to agree on how those inspections will be conducted. They must agree on: what information will be inspected; the schedules that will be followed, including any dates that are critical; who will be responsible for providing source material; who will participate in the inspections; and who will be responsible for reviewing information even though those responsible may not attend inspection meetings.
- The Outline Inspection: This inspection focuses on the outline (description of content and organization) as well as such other specification documents as the information objectives, information specifications, and audience and task analyses.
- The First-Draft Inspection: This inspection focuses on a first draft of the information—a draft that may be approximately 80 percent complete. Major technical errors and portions of missing information are defined,

Table 2. Collaborative Relationships During the
Six Phases of an Inspection

Phase	Writer Collaborates With	Activities
Planning phase	Moderator	Determine schedules and attenders
	Editor	Coordinate editing and inspection schedules
Distribution phase	Inspectors and Moderator	Ensure that all concerned parties receive copies and can review them
Preparation phase	Editor	Perform editorial rework while draft is being reviewed
	Moderator	Ensure that work is progressing on schedule
	Inspectors	Work on problems that may need to be resolved before inspection meeting
Inspection meeting phase	Moderator	Ensure that meeting is orderly and productive; resolve differences among inspectors
	Inspectors	Incorporate technical changes and additions
	Editor	Discuss editorial changes to draft; ensure that they do not cause technical change
Rework phase	Inspectors	Resolve unanswered questions
	Editor	Resolve editorial questions
	Moderator	Ensure that rework is progressing on schedule; obtain resolution on outstanding problems
Follow-up phase	Moderator	Report inspection results; schedule subsequent drafts and inspections

and plans are made for correcting the errors or omissions. At this inspection, the group of inspectors review and discuss the technical content and completeness of the draft to ensure that it agrees with the outline previously inspected and that the treatment of the material is, indeed, suitable for the intended audience.

- The Final-Draft Inspection: This is the inspection of the final (complete) draft. At this inspection, the inspectors examine the draft for the same reasons as for the first draft. In addition, they must ensure that comments from the inspection of the first draft have been incorporated.
- The Production-Draft Inspection: This inspection, which involves the writer and the production specialists, examines the draft that was submitted, after final approval, for final production work. The draft's adequacy is examined, and plans to supply final instructions to the printer (such as use of special paper stock or foldouts) are made. Schedules for production and printing are set as well as schedules for final input to the index (for example, final page numbers) and the last date that significant changes can be made without jeopardizing the production schedule.

The Role of Testing in the Information-Development Cycle

Like the other verification activities, the place of information testing in the information-development cycle is not restricted to one time and method; it can be done in different ways, at different points in the cycle, and to varying degrees.

The results obtained through information testing may vary depending on where in the information-development cycle the testing is done. For example, prototypes or information from earlier versions of the product or from related products can be used to test the effectiveness of the overall approach that the writer has taken when producing the information. Early or intermediate drafts of the information can be used to test specific functions of the product or specific attributes of the information. Final versions of the information can be used to test the actual use of the product and its associated information. Versions of the information that have been released to customers can be tested as a basis for producing the next version of the information.

Thing identifies seven places in the information-development cycle where testing can be done profitably [3]:

1. setting and testing objectives;
2. testing prototypes;
3. testing while writing;
4. task-oriented testing of early drafts;
5. testing against specifications;
6. task-oriented testing of the final draft; and
7. testing in the user environment.

Testing can, in fact, be done on any of the information items produced during the information-development process. When information objectives are set—very early in the process—they can be tested to ensure that they are reasonable and that the objectives being set are the ones that are important to measure. Similarly, prototypes developed early in the process can be tested to ensure that new approaches, graphic design, and information packaging are effective.

The testing can be done before the information is complete, while the information is being developed, when the information is completed but not yet made available, or after the information is in use by customers. Testing can be conducted as part of a comprehensive quality-measurement program that incorporates testing at many stages of the information's development. The testing may include use of checklists, peer reviews, laboratory testing, and customer surveys to measure and improve the accuracy, readability, retrievability, and task-supportiveness of information. Regardless of the methodologies used, it is most important that the testing involve a wide range of participants and that the participants work together to achieve a common goal. If the testing is not truly a collaborative effort, its focus may not be as broad as needed; it might, in fact, focus in on the preferences and views of a few individuals and thus fail to meet the needs of many potential users.

Generally, the earlier that testing can be conducted, the more useful and productive it may prove to be. Early testing and early emphasis on testing and test results can reduce the need for large testing and verification activities later in the product-development cycle.

The Role of Early Customer Involvement

Involving customers in verifying the information that they will eventually use is one way to obtain direct verification and feedback. However, this type of involvement is something that is usually considered after the fact, after the information has been produced, printed, distributed, and used. This need not be the case. Potential customers can be actively involved in all phases of verification— from ensuring that the requirements to be met are the right ones to ensuring that the information can be used successfully to meet those requirements.

Having examined the types of verification activities that are part of the information-development process, we now turn our attention to the collaborative activities that occur during information verification.

EDITING INFORMATION—TYPES OF EDITING

Three types of editing can be defined, although the distinctions among the types are not always clear: literary or stylistic editing, in which the editor examines the draft for correctness of grammar, wording, and punctuation; technical editing, in which the editor examines the draft for correctness and suitability; and copyediting, in which the editor examines the final version of the information for completeness, style, and adherence to standards shortly before the information is sent to the printer.

Editing always involves a close and intense collaboration between writer and editor, but the collaboration may well extend beyond them. Since one goal of editing is to ensure that writing produced by an organization presents a somewhat unified appearance—that of the organization rather than of the individuals—editing fosters a collaboration among individual writers, requiring them to work together, or at least confer with each other, to ensure a reasonable degree of consistency. Almost paradoxically, one contribution of an editor is to mask the collaborative nature of the information produced by an organization by giving it a unified appearance rather than the appearance of being the work of many collaborators.

Concerned primarily with the form and presentation of the information, the literary editor examines drafts for correctness and consistency of grammar, punctuation, word choice, style, format, and arrangement of ideas. While the literary editor focuses on the form and presentation of the information in drafts, the technical editor focuses attention on the content and use of the information. Although the technical editor is concerned with many of the same information characteristics as the literary editor, the focus of the editing is different—the technical editor must be concerned with the treatment of a technical topic to ensure that it is described thoroughly and accurately as well as appropriately and is stylistically correct. Unlike the literary editor who works with the material that is presented, the technical editor must frequently be concerned with what is *not* presented, uncovering gaps and omissions in the presentation and resolving discrepancies between books. Literary and technical editors must work together to ensure that they are not working at cross purposes and that they agree on standards and conventions to be followed. It is this all-encompassing view that distinguishes the technical edit from a technical review or inspection of the information; the technical editor can ignore *no* facet of the information, while technical reviewers can concentrate solely on the accuracy and completeness of the information.

Copyediting is most frequently done at the end of the information-development cycle, when final production work is being done so that the information can be sent to be printed. During the copyedit, the editor scrutinizes the information for adherence to the organization's style and formatting conventions, for any missing copy, or typographic errors. The copyedit is generally the last place to resolve inconsistencies and give the information a polished, unified appearance.

Editor-Writer Collaboration

Myers defines three types of relationships between writer and editor [4] :

- editor-centered relationships, in which the editor dominates;
- writer-centered relationships, in which the writer dominates; and
- team-centered relationships, in which editor and writer work cooperatively with each other and with others involved in the information-development process to produce the best possible product.

If the relationship centers too strongly on either the writer or the editor, they may come to see each other as adversaries, perhaps a reminder of "old newspaper days" where "tough" editors rode herd over new writers. Although there may be expedient reasons for having a short-term working relationship dominated by either the writer or editor (for example, if a seasoned editor is working with a group of inexperienced writers to produce a large volume of information on a very tight schedule), a more productive long-term relationship is built around true collaboration in which writer and editor learn from each other and work together to produce the best information possible. Editors must, in fact, be human-relation specialists to function effectively in their workings with writers.

Writers must learn how to work with editors in all phases of the writing process rather than just at the end so that the editor is a full contributor to the development of the information. This continuing involvement throughout the process saves time, produces better information units, and ensures a consistent, thorough treatment of the information.

One of the most important things that an editor can do is train writers to be better writers—in a sense, decreasing the need for editing of the writers' work. Applewhite points to the need for an editorial tutorial program in which the editor uses the writer's manuscript as a source for training exercises, tutors the writer in one-on-one sessions, and implants the improved forms directly back into the writer's paper [5]. Her training program has four phases: preventing errors, recognizing problems, establishing habits of effective presentation, and perpetuating standards of excellence in communication.

One major aspect of the collaboration between writer and editor is the editor's obligation to teach as well as to improve a writer's writing skills and abilities. Collaboration with an editor can be a very positive experience for a writer if the editor provides instruction, gives the writer the responsibility for understanding and making editorial changes, and motivates the writer to learn, grow, and improve.

REVIEWING AND INSPECTING INFORMATION

One of the most important information-verification activities is ensuring the accuracy and completeness of the information. This form of verification is done by giving the information to the technical experts (for example, those who are developing a product that the information describes) for their review and comment and then incorporating those comments into the information. While this procedure can be effective, it can also produce unsatisfactory results. For example, the technical experts responsible for reviewing the information generally have many job responsibilities related to their primary effort, which is most often the design and development of products to be offered in the marketplace. Reviewing information for technical accuracy is generally just one of many responsibilities and is often not seen as one of the most important ones. If time is limited, as it so often is, reviewing the information may be one of the tasks that is put off or done less thoroughly than it should be so that other tasks, ones deemed more important or more pressing, can be done on time. Similarly, technical experts may differ in their opinions of what is correct; they may thus return conflicting review comments about what is correct, leaving the writer to sort through them and make a determination. Human nature being what it is, the person who "yells the loudest" is often seen as being correct.

To remedy the problems inherent in information review, an information-inspection process, modeled on a successful programming-inspection process as described by Fagan [2], can be introduced as a replacement for the traditional review process. While the information-inspection process may seem to require more work to carry out (since it is, in fact, a thorough and structures process that places demands on participants), its long-term effect is to save time by reducing the amount of rework and last-minute change that would be required if a less efficient process for ensuring the adequacy of the information were used. The workings and advantages of this information-inspection process are described in this section.

Obtaining Reviews

With the traditional review procedure, a draft of a publication is produced and distributed to product developers for their review and comment. All areas of the product-development group should receive a chance to voice their comments and concerns, because their views, from their various vantage points within the project, may vary widely. When the reviewers have had a reasonable amount of time to examine the draft (about two weeks), they return marked-up drafts or descriptions of their comments and concerns to the writer, who incorporates their comments into the draft after resolving conflicts among reviewers' comments. Typical types of comments that a writer may expect are that pieces of information are missing, that technical changes have been made to the product being developed since the writer produced the information, that specifications have been misinterpreted, or that the reviewer does not like the specification and wants it changed. This last type of comment is usually one over which the writer has no control, but it is one that is frequently made; comments of this type can actually introduce technical inaccuracies into the information if they are interpreted as statements of fact rather than as dissenting opinions.

It may be wise to start the review process before a draft is complete by giving individual sections to technical reviewers piecemeal for their comments and suggestions. By using this approach, writers will not get too far ahead of the reviewers, and thus the amount of backtracking to incorporate comments, which otherwise may be extensive, is minimized [6].

But the burden of doing effective and thorough reviews should not be left entirely to the reviewers. Writers can do much to improve the efficiency and effectiveness of the information reviews. Writers must work to choose the required reviewers to ensure that all those who should review the information and comment on it are given their chance. To help those reviewers do the best job possible, the writer should state the objectives that each of these reviewers is to meet or the purpose of their review. The writer should tell the reviewers the type of comments and feedback being sought—detailed critique of the facts presented, verification of examples and illustrations, or approval of the general approach to the subject. The conscientious writer might go so far as to list the activities each type of reviewer must perform to ensure a thorough review. The writer must work with the editor before or after the review (or both before *and* after if schedules permit) to ensure that the information's editorial quality is ensured before it is distributed for review and after review comments have been incorporated.

By taking an active part in the review process and by working closely with the reviewers, writers can help to ensure that all reviewers are heard from, that their comments are understood and incorporated properly, and that reviews do not take so long that they jeopardize product schedules. Writers who ignore, or slight, the importance of proper review may be good writers, but they are not good information developers.

The Problems of Document Reviews

Proper and thorough reviews are vital to the success of a final document, but the traditional review process has inherent flaws that can make the process ineffective. Very frequently, writers must locate reviewers and coerce them to spend the time and energy to review documents. They must resolve conflicts among reviewers' comments and determine which ones are correct. Writers generally don't know how to judge which comments are correct and which aren't; as a result, reviewers who steadfastly support their positions have their comments incorporated. And, in most organizations, there is no effective way to pressure uncooperative reviewers to do their jobs.

Reviewers, too, have problems with the traditional review process. They generally don't know who else is reviewing the draft, so they don't know which areas to concentrate their review on and which ones can be reviewed less extensively because these areas are also being reviewed by someone else. Even in those instances when they know who else is reviewing the draft, they have no effective (or efficient) way to know what other reviewers have said about the draft. Based on the successful results obtained through use of the programming-inspection process an inspection process for publications was developed [2]. The traditional information review involved a distribution of drafts to reviewers, such as product developers. The writer would then incorporate these comments, which were sometimes contradictory, into a draft. In contrast, the inspection process described above relies on an inspection meeting for which participants would prepare ahead of time, in which they would participate actively, and after which they would perform required follow-up work as agreed to during the inspection meeting. The collaborative nature of the inspection meeting makes both the process and the product more effective.

Advantages of the Inspection Process
Over the Review Process

Because the information-inspection process is more clearly defined and enforced than previous review methods, technical experts charged with responsibility for inspecting the information are more involved in the process than they were when they were charged with responsibility as reviewers. Because those involved in the information-inspection work as a group, and the group's work is reported to all those responsible for developing the product, including the product's information, each person is made to feel more responsible for the quality of the inspection work. Hence the information receives a more complete review than it would have previously.

Inspectors at the inspection meeting benefit from group thought, so the review is a collective, consolidated one, rather than a number of isolated opinions. Reviewers also hear what other reviewers have to say. Because of the collaborative nature of the inspection process, the reviewers are responsible for resolving their differences; if they cannot reach agreement, one person—the person, termed the "moderator," who is assigned responsibility for conducting inspection meetings and ensuring that all those involved in the inspection process do their assigned tasks on schedule—is responsible for ensuring that agreement is reached and reported in a timely manner. The writer will receive one agreed-upon solution rather than having to decide among a number of alternatives.

TESTING INFORMATION

The information-inspection process detected defects in the information so that they could be corrected before the information was delivered to customers. The ability to detect and correct defects can be extended by introducing usability testing into the information-development process.

Testing information improves the product's interface to the customer by ensuring that the information is a vital part of the product, not a superficial feature, and that it contributes to the product's ease of use. Well-tested information can help a product gain acceptance by customers by reducing the customers' need for formal product education and by providing a good first impression of the product that will carry over into its subsequent use. By measuring characteristics of the information (either in a test laboratory or as part of early-product-version test), information developers can measure task performance including: time required to do a task, the number of errors made, and the amount of help needed, as well as the attitude of those taking the test towards the information and product being tested.

Atlas [7] and Soderston [8] have described information testing as another form, or level, of editing, one centered on users, but testing, in fact, adds a whole new dimension to information quality and the process used to achieve that quality. Testing ensures that the information is not only accurate, complete, and presented in the best possible manner, but ensures that it is suitable for its intended users to do the tasks that *they* want to do. It allows for additional improvements that cannot be obtained through traditional reviews or through information inspections. Usability testing is a way to determine, before a product is marketed, whether it is usable, and, if not, enables those responsible to make needed improvements before releasing the product to customers; thus customers need not wind up being "the first test subjects."

The major advantage of testing over other verification methods is that testing brings in viewpoints and perspectives about the information and the product itself, viewpoints that are not centered on product development, but on the product's users. These opinions cannot be obtained from those who were actively involved in developing the product. Testing can also bring a broader perspective to bear on the product and the associated information; it can be an effective way to learn if the product and information about how to use the product work well together in a user's environment.

PARTICIPATING IN EARLY SUPPORT PROGRAMS

There is a need for feedback on how a product and its associated information will be used that, on the one hand, reflects a more realistic use and audience than is possible in a laboratory test environment but, on the other hand, is available before the product is made generally available to customers. One solution to this dilemma is an "early support program" for the product, in which selected customers receive a version of the product before it is made generally available.

In an early support program, customers receive a version of a product to use some time before it is made generally available for customer use. These customers may receive special support for installing and trying out the product by having the product's developers on hand to offer advice and assistance and, at the same time, to determine how the customers use the product, what difficulties they may have in using it, and their overall satisfaction with the product. Results thus obtained may be used to modify the product before it is made generally available. This type of verification is different from other forms of verification and testing because it provides feedback from actual customers working in their own work environment. It is, in fact, not only an extension of testing, but also is an early version of releasing. By their very nature, early support programs must occur very late in the product-development cycle. The version of the product that is made available to those participating in the early support program must be in its almost final form. It must be a fully working version of the product with all the functional capability of the final product. Since the information for a product is developed on a schedule that is very similar to the schedule for the product, early support programs must occur very late in the information-development cycle.

Information developers may participate in early support programs by providing on-site support for early support program customers as part of the product-development group, by interviewing early support program customers to find out how they used the information provided and how well it met their needs, or by working with other product developers on telephone and written surveys of early support program customers to gauge their satisfaction. Working with other product developers they can determine what actions, if any, need be taken to modify the product and its associated information before it is released for general availability.

IMPLICATIONS OF VERIFICATION
AND COLLABORATION

One question of verification goals concerns the division of duties and responsibilities among the classes of verification activities. For example, it might generally be agreed that professional editors do the most effective and efficient job of editing. However, many product developers spend a large portion of their review effort on catching typographic errors or correcting word choice; this is usually not the best use of their time. Not only are they apt to be less qualified as editors than the editors are, but also their attempts at editing take away from the job they should be doing—reviewing the technical content of the information. The overall verification of the information may suffer as a result. On the other hand, it would probably be unwise for an information developer to bluntly tell a technical reviewer not to point out editorial problems with the information. The reviewer may be less willing to review the information when next asked. And who can say with absolute certainty that the editorial matters corrected by the reviewer would, in fact, be corrected during editing?

Similarly, certain types of corrections are better made during testing than during review, and vice versa. An accurate and effective way to ensure that each type of verification activity focuses on the appropriate characteristics of the information, which will in turn improve the overall quality of the information, is to plan verification activities as an integral part of information development and to ensure that all those who should be involved in verifying the information work together to do the best job possible.

Situational Dependence of Verification

It is often difficult to talk in general about verification because verification activities vary greatly with individual working situations. As a result, most analysis of verification activities focus on the individual instances of verification with less attention paid to generalizing procedures and practices that the importance of the topic warrants. Since verification is often done under pressure and on tight schedules, analyzing and documenting what has been done is often a low-priority item. As a result, much that is learned goes unreported and is not available for use the next time verification must be done. Consequently, many processes put in place to verify information are "built from scratch" rather than being constructed on a solid base established in earlier verification activities and benefiting from what was learned then.

Because the purpose of information verification is to correct and improve the information before it is delivered to customers, the goal of most verification programs is to be as effective as possible with less concern given to how efficient the verification programs are. With the increased emphasis on productivity, this is one area worthy of serious attention by researchers—to determine how to get the largest benefit for the resources expended.

The Collaborative Nature of Verification

Collaboration appears as an area of concern at each phase of the information-development process. Throughout the entire process, information developers must communicate with others to get the job done. For example, information developers must work with their technical contacts in the product-development organization, editors, test personnel, and possibly with customers in an early support program. The need for, and benefits of, collaboration appear most pronounced during verification, for this is truly the type of work that requires the skills, viewpoints, and expertise of many if it is to be done properly.

Some collaborative relationships show up on the organization chart; they are the formal, often permanent, ones. Others are more transient—individuals or groups working together to solve a specific problem or group of problems. If people within an organization rely solely on the formal relationships for their collaborative activities, they may not be able to respond in the best possible manner to the immediate needs of the organization. People must be willing and able to collaborate with those who are best able to supply the talents and skills needed to solve the problems at hand. Collaboration brings together a variety of skills and viewpoints; it is a way of bringing together all the concerns the verification must satisfy and using the combined skills of the collaborators to find the best solutions to the problems that are raised by those concerns. The scope of collaboration must be defined to be as broad as possible. Potential collaborators can be found within an organization, but the best people with whom to collaborate may well be outside of the organization, for example, collaboration between industries and universities, collaboration with customers or potential customers, or collaboration with outside consultants who serve as experts in their fields.

Unlike many other information-development activities, verification is distributed throughout the entire information-development process. Verification is not an activity that is done once; it is done during each phase of information development—from defining information objectives through producing camera-ready copy—to ensure the quality of the output from each information-development phase. Throughout the information-development cycle, writers collaborate with many others in their organization to ensure that the information they produce is accurate and complete, usable by customers who must perform specific tasks, suitable for its intended audience, and presented clearly and with style. Each of the people with whom the writer collaborates to verify the information and ensure its quality brings specific skills and viewpoints to the verification activity, ensuring that the information is verified as well as is possible.

Verification during product development cannot uncover all defects. Actual use of the product by customers to do their jobs, and the effects of learning and product familiarity on use, cannot be assessed until the product has been in use for some time by customers. Despite these limitations, time and effort spent in doing effective verification are well spent. Verification is an important part—an integral part—of any information development process. Whether the verification is done in a formal, well regulated fashion or more informally, because of restraints on time or facilities, it requires the collaboration of a variety of an organization's personnel—writers, editors, technical experts, graphic designers, and usability experts—and sometimes people from outside the organization. Time and resources must be built into schedules and resource plans during the early planning phases of the information-development process and the overall product-development process. People with needed skills must be made available to work on information verification. If the time and resources are not planned for from the very start, it may be difficult, if not impossible, to add them later.

REFERENCES

1. C. B. Wagner, Quality Control Methods for IBM Computer Manuals, *Journal of Technical Writing and Communication, 10*:2, Baywood Publishing Co., Inc., Amityville, New York, pp. 93-102, 1980.
2. M. E. Fagan, Design and Code Inspections to Reduce Errors in Program Development, *IBM Systems Journal* 1976 No. 3, International Business Machines Corp., Armonk, New York, pp. 182-211, 1976.

3. A. L. Thing, The Test-Driven Information Developer, *Proceedings of 31st International Technical Communication Conference,* Seattle, Washington (April 29–May 2), Society for Technical Communication, MPD-34-MPD-37, 1984.

4. B. Y. Myers, A Classification of Author-Editor Relationships: Toward Team-Centered Relationships, *Proceedings of 31st International Technical Communication Conference,* Seattle, Washington, Society for Technical Communication, Washington, D.C., WE-116-WE-119, 1984.

5. L. B. Applewhite, An Individual Development Program, *Proceedings of 29th International Technical Communication Conference,* Boston, Massachusetts, Society for Technical Communication, Washington, D.C., E-14-E-16, 1982.

6. J. J. Vreeland, GET IT WRONG THE FIRST TIME–So You Can Get it Right by First Draft, *USER-bility Annual Symposium Proceedings* (July 31–August 1, 1984, Poughkeepsie, New York). International Business Machines Corporation, Kingston, New York, pp. 5–8, 1984.

7. M. A. Atlas, The User Edit: Making Manuals Easier to Use, *IEEE Transactions on Professional Communication* PC-24, No. 1, The Institute of Electrical and Electronics Engineers, Inc., New York, pp. 28–29, 1981.

8. C. Soderston, The Usability Edit: A New Level, *Technical Communication,* First Quarter, Society for Technical Communication, Washington, D.C., pp. 16–18, 1985.

CHAPTER 12

Collaborative Editing: A Combination of Peer and Hierarchical Editing Techniques

HENRIETTA NICKELS SHIRK

Editing is a collaborative act. Whether it involves persons other than authors scrutinizing documents or authors themselves in roles of editors of other authors' documents, it is a collaboration of one mindset with another.

Technical editing has been aptly presented in definitions such as that by Lola Zook, who describes it as "taking a manuscript that presents factual or conceptual material about specialized subject matter, and doing whatever is necessary to prepare that manuscript for effective communication with the intended audience" [1]. Editing may be considered a profession, a job, a skill or capability, a set of techniques, and even an art.

While most communication professionals would agree on the activities and processes involved in editing, answers vary greatly to the question of who most effectively performs these actions. This chapter examines answers to this question of the appropriate location of different editorial responsibilities by looking at editorial collaboration by individuals within both academic and industrial settings. It reports the results of a study of authors and editors and their reactions to editing the work of others and of being edited themselves.

The purpose of undertaking this research project was to accomplish the following objectives:

1. identify the strengths and weaknesses of peer and hierarchical editing in both academia and industrial settings;
2. describe the rationales and organizational environments in which peer and hierarchical editing endeavors each tend to flourish;
3. observe whether peer or hierarchical editing is more successful in organizational settings; and
4. determine what skills should be taught in both industry and academia to ensure the success of technical editors.

For the purpose of this study, the roles of editors are described from two perspectives—peer and hierarchical. Peer editing is defined as any activity performed by an author's organizational equal for the purpose of improving or assisting a document toward its publication. These peers may be fellow writers functioning as temporary editors or professional editors at the same level as the writer in the organizational hierarchy. On the other hand, hierarchical editing refers to these same activities performed by an individual higher in the organizational structure than the author (such as a manager, an instructor, or a professional editor).

The study reveals both the strengths and weaknesses of the two approaches to the editing task. Based on these results, several conclusions are drawn concerning the typical organizational structures in which peer and hierarchical editing each tend to flourish. Finally, the pedagogical implications of the study suggest several revisions to current methods of training technical editors and the implementation of innovations to promote a new perspective called "collaborative editing," which is a synthesis of both peer and hierarchical editing.

SAMPLE POPULATIONS

The data from academia for this study were gathered over a three-year period during which I taught technical and professional writing courses at Northeastern University, Boston, Massachusetts. Students in my upper-division courses in "Technical Writing" and those in my graduate-level courses in "Technical Writing" and "Technical and Scientific Editing" were the sources of information regarding their editing experiences in the classroom setting. One hundred students participated in the study, fifty undergraduates and fifty graduates.

The data from industry for this study was gathered during the same period of time by me and by graduate students in my "Publications Management" course at Northeastern during spring quarter 1988.[1] We contacted 100 technical writers and editors currently working in a wide variety of organizations in business settings throughout the greater Boston locale.

Additionally, my graduate students and I relied upon our own personal experiences in the field of technical communication. My own background includes fifteen years of experience as a writer/editor and publications manager in the software industry, during which I established editing functions at three major corporations which previously had no employees working officially in an editorial capacity. The majority of my graduate students were either currently employed as technical writers or editors in industry, or they had recent relevant experience in these capacities. This experiential data were used to supplement the information gained from the 200 participants.

HOW THE STUDY WAS CONDUCTED

The research for this study was conducted using both quantitative and qualitative methods of investigation. This combination of approaches seemed appropriate because editing is both an objective and a subjective act of collaboration, often involving the application of generally accepted writing and graphics standards within the framework of an intensely personal work relationship between writer and editor.

Questionnaires were distributed to the students in the sample, asking them to rate their responses to a series of questions about the quality of their editing experiences in the courses they had just completed (see Appendix A). This questionnaire also gathered data on the students' prior experiences with both peer and hierarchical editing. Finally, it included some open-ended essay questions designed to elicit responses regarding the emotional and interpersonal aspects involved in the collaborative acts of editing and being edited.

Questionnaires were also distributed to professional writers and editors in several organizational environments. Because these groups varied so widely in their makeup and assignment, the questionnaires were modified somewhat for each group. However, the questionnaires all collected data about the structure of each organization and the respondents' relation and reaction to the editing processes within it. These questionnaires also included open-ended questions regarding the quality of the respondents' editing experiences (see Appendix B for typical questionnaires).

[1] I am grateful to the graduate students in my course in "Publications Management" at Northeastern University during Spring Quarter 1988 for their assistance in accomplishing the research for this essay and for their sharing of personal experiences in the field of professional communication. These students are: Lori Alexander, Elena Aschkenasi, Lori Caldwell, Mary-Ellen Finlay, Christine Kelly, Cecelia Lee, Martha Lillie, Anita Mateyak, Melissa Murphy, Amy Schoonmaker, Eric Schuster, Mark Shea, Nancy Veiga, James Wolstenholme, and Annie Wynn.

RESULTS OF THE STUDY

Academia

Because students at Northeastern University had experienced hierarchical editing from writing instructors in traditional composition classes, the questionnaires distributed to them focused on what was a new experience for the students—peer editing. Of the 100 students surveyed by the questionnaire, 14 percent felt that they did exceptionally well on the writing assignments for which they had peer editing. Eighty-two percent believed that they did better than expected, and 4 percent believed that the peer editing made no difference to their performance. No one felt that they did worse than expected due to the edit. A vast majority of the students felt they benefited from peer editing.

As a control measure, some of the assignments in the courses surveyed were not peer edited. Students responded to the question of whether they would have liked to have had all their assignments for the course edited by a peer with an impressive 97 percent in favor. Two percent were not in favor of additional peer edits, and 1 percent had no opinion. The activity was obviously considered successful.

The question relating to the primary contributions of peer editors to writing assignments revealed some differences between the undergraduate and graduate students (see Table 1). The numerical differences suggest that undergraduates seek help in the mechanics of writing, while graduate students tend to look at larger conceptual issues relating to organization and structure.

The responses to the question about the experience of each of the students with peer editing also revealed some significant differences between undergraduates and graduates. While 15 percent of the undergraduates and 23 percent

Table 1. Editorial Contributions of Peer Editors

Editorial contribution	Percent of undergraduates	Percent of graduates	Percent of two groups
No contribution	0	0	0
General organization	13	28	21
Word choice (style)	15	35	26
Correct grammar	42	8	25
Correct spelling	34	3	19
Correct punctuation	26	5	16
Sentence structure	17	11	15
Paragraph structure	9	17	13
Other (content, motivation, etc.)	3	2	3

of the graduates had received a peer edit from a classmate, the figures for peer editing from colleagues and non-teachers had a wider spread. These responses showed that only 29 percent of the undergraduates had received a peer edit from a colleague at work, compared to 61 percent of the graduates. The diversity of responses was also similar for the non-peer edits (from non-teachers such as supervisors), with 50 percent of the undergraduates as compared to 86 percent of the graduates responding in the affirmative.

The fact that most of the graduate students at Northeastern are already employed in various professional communication positions in business and industry may account for some of the differences between the two groups. The graduate students were obviously more similar on a general skill level to the group of professional writers and editors surveyed in industry. The graduate students therefore edited at a more professional level.

While the measurable aspects of the student questionnaire were fairly predictable, the comments in response to the open-ended questions revealed some interesting aspects of the editing process and its impact on the writers and editors involved. These covered the four areas of audience, setting, psychology, and problems as they relate to editing.

The most frequently mentioned benefit of peer editing was the fact that the process gave the students an "audience" for their writing other than the instructor. For many of them, this was the first time they had obtained such feedback on writing assignments. The following comments were typical: "It enabled me to see a view of what others saw when they read my papers," and "The peer editing was good for an objective view of my work; it enabled me to have an *audience.*"

The fact that most technical writing involves a consideration of multiple audience levels was highlighted by the fact that the students usually experienced different editors at various points in the course. Responses to this multiplicity of editors reinforced the importance of being sensitive to the needs of different audiences. As one student revealed: "It is sometimes confusing when two different editors pick up different things. One editor will indicate something for improvement, and the other editor will consider it fine. I guess this means that there is a variation in audience." And another student observed: "I realized how others would perceive my comments. I had two editors, and each had unique comments and corrections."

The notion of a real audience with multiple levels of comprehension of one's subject matter was also emphasized in the reactions to specialized terminology and organizational issues. Some of the in-class editing work during which students read their papers to small groups of classmates produced these reactions: "I was able to read my paper to people who are not computer science majors and who would not understand some of my terminology." And from several students: "I had other people listening to me that did not know much about my topic." Students who commented about changes resulting from having these different perspectives on their writing, generally agreed that they were more likely to "define terms that were unfamiliar to the uninitiated and to include points that I missed or didn't think of." Differing perspectives served as catalysts for revisions.

The importance of the atmosphere in which peer editing takes place was emphasized by several comments about the classroom setting. As one student observed: "It always helps to have someone else read your rough draft, especially when it is in a non-competitive, supportive atmosphere." Other students reacted positively to their editing experiences, because they saw the peer feedback they received as helpful criticism. "I think it is valuable in any setting," explained one student, "to get another, non-threatening, point of view." Another student, who admitted to being "hypersensitive," found that the supportive environment of the classroom editing situation was extremely positive, because "it taught me how to deal with criticism better than I ever had before."

The reaction to criticism was also the basis for many comments relating to the psychological issues of the editing process. It is interesting that most of the students commented at greater length about what it felt like to be edited, rather than about their experiences of being an editor. However, the language used to describe these experiences, was generally both positive and negative. Peer editing was called "a bittersweet experience," "a mixed blessing," and "encouraging as well as discouraging." One graduate student summarized this widespread feeling of ambivalence as follows: "I reacted to being edited in two ways—with both sides of my brain. Emotionally, being edited did not feel good, because my right hemisphere perceived it as a personal attack. Not having worked with an editor before, my left hemisphere found it difficult trying to control the 'Fight or Flight' syndrome. Rationally, I knew not only that the person who edited my work had nothing against me, but also that being edited would probably improve what I had written." Although most students recognized the benefits to be derived from peer editing, they generally concluded that it would take more time and experience to feel comfortable with it.

One of the problem areas for students was that of giving criticism. Although some felt that it was "gratifying" to work on someone else's writing and "interesting and helpful to see how someone else organized their paper," most were uncomfortable about criticizing the writing of classmates. "I felt nervous about pointing out the bad things," admitted one student, "because I didn't know the author that well and didn't want to hurt her feelings." This hesitancy was also reflected in the comments of authors about being edited. From an undergraduate: "Generally, the editors were not critical enough about the material they were editing. They made small changes in grammar, but not in other areas such as organization." And from a graduate student: "There was a reluctance of fellow students to be totally honest in identifying problems. I think this is based on the fear that if you find problems the other person is going to take it personally instead of as constructive criticism." About half of all the students wished that their peer editors had been more critical.

Two additional problem areas emerged in the student comments. First, some of the students were concerned about inequalities in the writing abilities of paired writers and editors. A few students suggested matching writers and editors in terms of writing skills, asking for "an equal level of vocabulary and grammatical understanding." This request was supported by others who "couldn't get good feedback, because English was a second language for their editors." Second, some of the students could not escape the hierarchical aspects of their previous educational experiences in writing. "I would have rather have had you (the instructor) edit my work," admonished one student. And another student confessed: "I would have preferred a teaching assistant to edit the paper, because I would *know* whether I was wrong, or whether it was a personal choice." However, only six percent of the students identified these two areas as being negative aspects of their experience.

Generally, the remaining 94 percent of the students were positive about their peer editing experiences. Even those students who chose not to implement their peer editors' suggestions recognized the value of the experience. "Although I did not always use my peer editor's advice," admitted one student, "I did find it helpful to get an outsider's opinion." And another student reacted with: "I appreciated the comments that were made and found them to be, in most cases, helpful. Although I have chosen to ignore a few of the suggestions made by my editor, this is a choice I can make with a better awareness." The awareness value of peer editing was aptly summarized by another student who felt that this activity provided the "Aha!" of the course, enabling him to more clearly understand areas for improvement in his own writing.

Many of these views of editing in the classroom were confirmed by the research conducted in industrial organizations. However, this part of the research also revealed some additional dimensions to the collaboration involved in the act of editing.

Industry

The data collected from organizational settings in industry covered several broad topics. Because of the slightly varying versions of the questionnaires administered to the participants, for the purposes of this discussion we will only examine some of their significant areas of agreement across industries. As in academia, peer and hierarchical editing have both strengths and weaknesses. The respondents in this study identified these elements, and their responses suggest the rationales behind each type of editing. Of the 100 participants (including both editors and writers), about one third each worked in peer, hierarchical, or a combination of peer/hierarchical editing settings. This equal distribution assured that the responses did not favor any group.

Strengths—better quality documents — Writers' and editors' perceptions about the benefits of both kinds of editing revolved around questions relating to the perceived relationships of writers and editors and their views of the quality of the documents that were subjected to each kind of editing process.

In one large computer company in which both peer editing and hierarchical editing by professional editors occur, 80 percent of the editors believed that their edits provided a more disciplined reading of documents than that provided by peer editors. However, when asked whether a preliminary peer review improved the final quality of publications, 60 percent of these same editors responded that it did, and no one disagreed with the question.

Likewise, when writers in this same company were asked if peer reviewing improved their own writing ability, 82 percent agreed that it did. They also felt (64%) that the peer review process improved the quality of documents. However, this same group of writers believed that professional editors fulfill a necessary role in their documentation process. Most agreed (64%) that editors are more effective than peer reviewers in the overall improvement of a document. They also agreed (77%) that their own writing had improved because of the advice received from professional editors.

These varying responses initially appear to be ambivalent reactions to the two kinds of editing. However, they indicate an appreciation for the benefits to be derived from both kinds of editing and point to the inherent strengths of each.

Peer editing was generally perceived as less threatening than hierarchical editing, with 89 percent of those surveyed indicating that they seldom or never felt threatened by the peer editing experience. The majority of the respondents (83%) felt confident when a peer edited their work, and 94 percent of the writers surveyed seriously considered the changes proposed by a peer editor for incorporation into the final versions of their documents. These findings parallel those of the students in Northeastern's classrooms, who generally viewed peer editing as a positive, non-threatening experience in which they learned how to improve their writing.

Peer editing provides the obvious benefit of improving the quality of documents (according to 60% of the editors and 64% of the writers in this study). When the industrial writers were asked if they devoted more time and energy to documents that went through a peer review as opposed to those that did not, 77 percent responded affirmatively. They felt that they invested quality time to produce documents which would be edited by their peers, exerting more effort and accepting the criticism they received. Writers appear to become more aware of their roles as writers when they must deal with their peers. In this way, peer editing is similar to peer pressure but without the negative connotations. For writers, the fact that their work will be edited by a peer causes them to perform beyond their usual expectations.

Likewise, hierarchical editing was perceived to have several strengths. The survey responses in this area were similar for hierarchical editing that was performed by a professional editor higher in the organizational structure than the writer and that was executed by a supervisor or manager. In either case, the hierarchical editor was perceived to have more credibility than the peer editor. In one segment of the study, 61 percent of the respondents believed that hierarchical editing was more valuable than peer editing. Reasons for this perspective were that the hierarchical editor knows the "big picture" of the organization's publications and can assure a unified and consistent image in terms of writing style and visual presentation. Hierarchical editors were often perceived, by the very nature of their position in the organizational structure, to have more qualifications to perform editing tasks. These respondents were similar to the students who preferred to have editorial feedback from their instructor, because they considered it more credible.

Weaknesses—differing personalities and experience — Often what is considered a strength or a weakness by one person in a particular setting can be just the opposite to another person in the same or a different setting. The limitations of peer and hierarchical editing may be affected by the type of organization and by the personalities and experience of the writers and editors involved.

Like their counterparts in the classroom, professional writers generally felt that peer editing was both non-threatening and at the same time threatening. When a peer critiques a peer's work, the action itself seems to create an inequality in the relationship of the writer and writer-as-editor that many find creates personal tension. In one segment of the study, this limitation of peer editing was underscored by the fact that 55 percent of the respondents believed that a peer editor usually made suggestions that reflected personal preferences rather than improvements to the document. Issues of perceived "ownership" of documents and authors' egos also emerge from this ambivalence toward editing. Authors like the feedback from editors because it improves their writing and enhances their professionalism, but the process of obtaining this feedback is often perceived as emotionally painful.

Another issue that arose in discussions with peer editors and writers whose work was reviewed by them was a concern for lack of time. As one writer-editor confessed: "When I'm working on my own deadline, I don't spend as much time as I should reviewing a peer's work." The quality of peer edits may therefore fluctuate greatly, depending on scheduling issues. However, this same lack of time could become a strength when it forces focused attention to the editing of a manuscript within a short time period.

Likewise, hierarchical editing's strengths can also be perceived as its weaknesses. The major problem is that the hierarchical editor may misapply the authority vested in the position by insisting upon personal preferences that may actually weaken the effectiveness of the document. One writer explained that he once left a writing position because his manager/editor insisted that he write in the passive voice—something his standard of professionalism could not tolerate. In one segment of this study, nearly half (44%) of the respondents said that they frequently felt threatened by a hierarchical edit, while only 11 percent admitted to feeling threatened by a peer edit. Although the hierarchical editor is usually in a position to enforce standards and therefore consistency, editors should assure that these standards are reasonable and that they are implemented in a non-threatening style and environment.

Some of the open-ended questions on the questionnaires revealed further concerns about how hierarchical editing was implemented. Lack of interpersonal communication skills was the most prevalent of these concerns. As one writer justifiably complained: "I write my manual and give it to an editor; then I never see it again until it is published." This individual was obviously in need of appropriate feedback. Clearly, the atmosphere in which edits are performed are an important factor for the success of the collaborative method selected.

The importance of communication styles and editorial environments receives attention in the following discussion, which summarizes the statistical results of this study and their implications.

DISCUSSION OF RESULTS

The perspectives on editing provided by this study point to several different attitudes regarding the collaborative process among writers and editors. These have significant implications for both organizational and pedagogical environments.

Organizational Implications

When combined with demographic data concerning the respondents' job responsibilities, experience levels, and organizational structures, both peer and hierarchical editing point toward several typical organization characteristics (see Table 2).

Table 2. Characteristics of Editing Environments

Organizational element	Peer editing	Hierarchical editing
Size of publications group	Typically small	Typically large
Work atmosphere	More informal	More formal
Job responsibilities	Multiple roles per person	Well-defined and specialized roles
Types of publications	Short documents	Long documents
Editing standards	May not exist, or relaxed application	Usually exist with strict application
Management style	More democratic	More authoritarian

While Table 2 reflects only general tendencies, it does identify the directions toward which certain kinds of organizations with specific publication needs tend to gravitate. Peer editing is more likely to flourish in small, unstructured environments where shorter documents such as brochures and newsletters are written. On the other hand, hierarchical editing is more likely to flourish in large, more structured environments where longer documents such as manuals and proposals are written. From a management perspective focused on the goal of producing high quality documents, there are advantages in implementing both approaches to editing.

Recognizing the strengths and weaknesses of both kinds of editing, the question becomes how to best manage the two processes. Hierarchical editing is obviously managed within an organization's existing structure. However, the issue of managing peer editing was addressed by the study's respondents, who were equally divided as to whether the process should be allowed to evolve informally among peers as an aspect of the writing cycle, or whether the process should be directed by a supervisor. The "correct" answer to this question will depend on the corporate cultures and the management styles of those involved. In many writing groups, peer editing will continue to occur whether or not it is officially approved by management.

However, managers can encourage the attainment of high-quality documents by assuring that there are appropriate publications standards in place. Standards result in a consistent, rather than a haphazard and individualized, approach to editing, regardless of who collaborates on the edit. The importance of standards was emphasized by one segment of this study in which respondents in both peer and hierarchical settings indicated that the major weakness of the current editing system in their organizations was the lack of formal writing and editing standards. Without the direction provided by standards, publications will tend to be reflections of individual preferences rather than of a consistent corporate image.

The data from this study also suggest that writers and editors in both industry and in academic settings experience a certain amount of psychological stress and ambivalence in the act of editorial collaboration. This is true for both hierarchical and peer editing. Although this stress tends to lessen with maturity and the experience of editing and being edited, it is nonetheless a persistent factor in this kind of collaboration. The issue seems to revolve around the perception of threat in giving and receiving criticism.

Some difficulties in this area are undoubtedly due to our societal attitudes toward the act of writing. Our educational system fosters the notion that writing is a private activity which produces a product that one "owns" and views as a kind of extension of one's self, which is then submitted for scrutiny by others. While this personal aspect of writing aptly applies to what we call "creative" or "nontechnical" writing, it is not appropriate for "technical" writing.

Writers and editors in industrial organizational settings must view most of their writing as an attempt at objective communication, not as an extension of the self. Written products in such settings are public representations of the organization, not private extensions of the individuals involved in their creation (although some internal documents, such as memos, may express private opinions intended for only a limited audience). Collaboration, whether it be in hierarchical or peer editing contexts, places the processes and products of communication in a public arena. Often, technical writers' and editors' names do not even appear on their work, because the work represents the organization, not the individuals whose efforts created it.

This "ego-less" approach to writing and editing can best be fostered in an organizational environment where management is aware of the benefits of both approaches to editing and does not allow hierarchical editing to interfere with peer editing or vice versa. Establishing a collaborative atmosphere is a management responsibility central to successful editing. This environmental ambience should be open and flexible on a psychological level, while fostering the concept of a "community" of collaborators. Tolerance for and encouragement of both approaches, as well as knowing when to best apply each, are crucial elements.

Time is always a crucial factor in the management of organizational resources. Given that there is always a shortage of time in meeting writing and editing schedules, a successful combination of both kinds of editing can help to relieve some of the stress of publication deadlines. This is especially true in organizations which produce different kinds of documents. Depending on the complexity or length of the documents, they may require only one of the types of edits. In other instances, both editing approaches will complement each other. Knowing how to orchestrate the two collaborative processes can become the measure of a manager's and an organization's success.

For example, when deadlines are tight, a publications manager could assign specific types of edits for the same document to different individuals. A hierarchical copy editor could be asked to examine a document only for grammar, spelling, and punctuation. At the same time, a knowledgeable peer of the author could be checking the document for accuracy of content. And, finally, the manager could assure that the document meets corporate legal and format requirements. However such diverse tasks are assigned, an important aspect of the collaborative editing process is that many facets occur simultaneously and as part of a team effort involving both peer and hierarchical elements.

Although demonstrated in a corporate environment, the collaborative approach described above is not different from that proposed for classroom situations by Peter Elbow in *Writing Without Teachers*. In looking at the psychology of writing in small groups, Elbow recommends that teachers of writing classes become participants in the roles of learners and thereby through example encourage the proper environment in which collaboration can occur [2]. Likewise, the corporate publications manager can provide opportunities for collaborative editing through fostering team efforts involving both peer and hierarchical approaches.

Pedagogical Implications

Given the results of this study and their implications for organizations, the question remains as to how best to prepare writers and editors for the collaboration required by their work. Such preparation may be within the academic setting of technical writing and editing courses or within training sessions in organizational settings. Both educational arenas have identical requirements.

These requirements were identified through the questionnaires administered in this study which contained a section asking practicing communications professionals to identify the skills they believed most needed to perform effectively on their jobs (see Table 3 for the general ranking of these skills by both writers and editors).

That editorial skills should be developed in those who pursue an editing career is not surprising. It is notable that both editors and writers agree in placing strong emphasis on diplomacy and interpersonal communication skills and less importance on the technical content. This data parallels that gathered in the classroom, where students expressed more concern about the psychological and interpersonal aspects of the editing process than about their understanding of language usage and accuracy of content in the documents they edited for peers.

In many ways, the classroom environment is easier to change than many business environments where social customs and channels of communication may already be firmly established. However, like the manager, an instructor is responsible for creating the ambience and "rules" of the editing situation. Both peer and hierarchical editing can be effectively combined in the classroom, with students and instructor working as partners toward the goal of creating documents that communicate effectively. Several excellent descriptions of assignments for accomplishing this task have been presented elsewhere [3, 4].

Table 3. Skills Necessary for Effective Performance

Skill required for success	Editors' ranking	Writers' ranking
Editorial Abilities (including grammar, spelling, punctuation, organization, style, etc.)	1	3
Technical Knowledge (of content, product, or industry)	4	4
Diplomacy (tact, respect, patience, courtesy, and perseverance)	3	2
Interpersonal Communication Skills (ability to get along with others through well-developed listening and interviewing techniques)	2	1

Several essays on collaborative learning by composition theorist Kenneth A. Bruffee may be useful when planning to teach collaborative editing skills within classroom contexts. Bruffee not only provides a brief historical background regarding collaborative learning [5], but he also describes some practical models for implementing it in writing classrooms [6]. Bruffee reassuringly points out that "what distinguished collaborative learning in each of its several types from traditional classroom practice was that it did not seem to change what people learned ... so much as it changed the social context in which they learned it" [5, p. 638]. Not only did students' work improve through the assistance of peers, but the peer helpers themselves learned from the students they assisted and from the activity of helping itself. Collaborative learning, according to Bruffee, can be a very powerful educative force, and it can be a means of acquiring new knowledge.

In explaining how to write collaboratively, Bruffee maintains that learning to write is different from learning anything else. It is not learning facts, but learning to do something. Learning to write, he suggests, "is in great measure a process of gaining new awareness. Gaining new awareness of any kind is likely to be a painful process. People need some kind of support while undergoing it. And the evidence provided by collaborative activity in the society at large suggests that people can gain both awareness and support as adequately in a small group of their peers, as from the ministrations of a teacher" [6, p. 640]. This observation also describes editing.

The question remains, however, concerning *how* one should best approach teaching the psychological and emotional aspects of the editing experience—how one should deal with the pain of increasing awareness. The concept of "ego-less" writing is often so foreign to students who have been in traditional educational environments where they work only for their own grade that they are initially hostile to spending personal time editing someone else's draft. It is only when they begin to see the benefits that accrue to their own writing that they recognize the value of collaboration. This initial barrier is the easiest to overcome, because it readily becomes evident through practice. More difficult are the barriers relating to interpersonal communication.

By definition, good editing is not a solitary activity. It must be accomplished with other people with whom one shares a sense of comradery and collaboration. The ability to work collaboratively rather than competitively does not come easily, but it can be encouraged through using several concepts from the field of management communication. The materials and studies which relate to styles of communicators are particularly helpful.

A commonly accepted general definition of communicator style is: "The way one verbally and paraverbally interacts to signal how literal meaning should be taken, interpreted, filtered, or understood" [7]. Understanding of differing communication styles and the perceptions that result from the words and actions of others is crucial to any successful editing process.

There are numerous techniques for assessing communication styles. Although the Myers-Briggs Type Indicator (MBTI) based on Jung's theory of personality types is frequently used in technical communications courses, there are several other instruments which have been developed to measure communicator styles. Several of these are described and evaluated in a recent article in the *Management Communication Quarterly*, which suggests that it might be wise to administer more than one of the instruments [8]. Whatever method one uses to focus on styles of communication, it is important to consider the context in which editing normally occurs.

Editing involves some degree of stress, regardless of one's level of experience. Understanding varying communication styles and non-verbal behaviors provides students with a conceptual framework within which to understand individual communication differences. It also provides them with the vocabulary to discuss their own feelings and reactions to the styles of others as well as to their own. Building an appreciation for these differences and for how communication styles may change somewhat within the stressful context of the act of editing is important for successfully teaching editing skills.

The advantages of writing for and giving feedback to peers in the classroom workshop setting has been well documented in terms of teaching the importance of audience, organization, and language use [9]. However, the workshop concept needs to be expanded within such courses to include coverage of the psychological and emotional aspects of editing. Administering a communication style inventory and providing some follow-up discussion on the results are helpful activities, but they are not in themselves sufficient for enabling students to understand the full impact of communication style differences. Workshops should be created in which students role-play various editing situations in terms of their own and differing communication styles. It is only by experiencing these differences within a reassuring workshop atmosphere, that students can begin to overcome some of the barriers to effective editorial dialogue.

The study of communication styles must be an integral aspect of teaching editing. It is the most effective mechanism for developing the collaborative skills required by editing.

CONCLUSION

Both peer and hierarchical editing are employed with equal frequency and equal success in industry. The strengths and weaknesses of these two collaborative approaches to the task of editing are confirmed by practices in both academia and industrial settings. While the environments in which editing occurs are often an influence upon the kind of editing performed (especially in terms of time and human resources), they can also provide opportunities for change and expansion.

As a collaborative act within organizations and classrooms, editing is also a public act in which responsibility for the success of written products is shared.

Not only does the writer relinquish some individual control over the document to the editor, but both in turn transfer some of their ownership to the group effort or the organization itself. Editing does not confer ownership of the document, but rather results in collaboration which contributes to the success of the group or corporate entity. Like many public events, editing can best be served by including all those who are best able to contribute to its success (both peers of writers and those higher in organizational hierarchies). The most effective means of accomplishing this success is through "collaborative" editing—a combination of both peer and hierarchical editing techniques in the service of effective communication (see Figure 1).

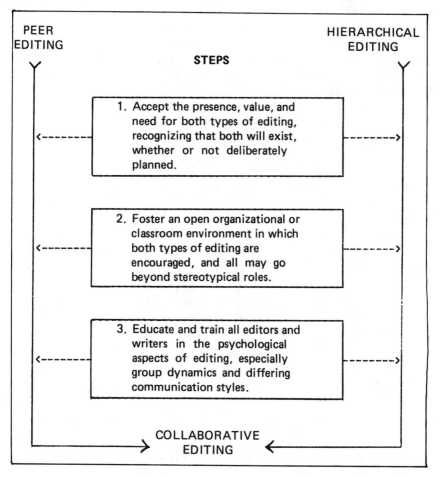

Figure 1. How to achieve collaborative editing.

This study of peer and hierarchical editing in academia and industry provides evidence for the importance of both approaches to editing. By identifying the strengths and weaknesses of these two approaches, one can see how they actually complement each other and truly provide a "community of collaborators." Most importantly, this study reveals the significance of organizational and academic environments for the success of editing. Without attention to the psychological and emotional factors, any kind of editing possesses the potential for failure. Those learning to become writers and editors must be placed in situations where they can learn and practice the theories of styles of communication as well as practical editing techniques in responding to the challenge of improving written documents.

By following these recommendations, professional communicators will not only develop confidence in their own abilities to edit, but also the appropriate levels of emotional empathy to assure the success of the collaborative effort. Both peer and hierarchical editing are centered on improving the document. The additional dimensions suggested here assure that editing is also centered on improving the relationships among the people who create the document. Collaboration requires attention to both dimensions.

APPENDIX A:
Questionnaire Distributed to Students at Northeastern University

Peer Editing Evaluation

1. How do you feel that you performed on the class assignments for which you had a peer edit your work?

    Exceptionally well: _____
    Better than expected: _____
    Made no difference: _____
    Worse than expected: _____

2. Would you have liked to have had a peer edit for *all* your assignments in this course?

    Yes: _____
    No: _____
    No Opinion: _____

3. What were the *primary* contributions of your peer editor(s) to your writing assignments?

    [Check all items that apply and then rank them in priority order with #1 being the area most important to you.]

    _____ No contribution
    _____ General organization
    _____ Word choice (style)
    _____ Correct grammar
    _____ Correct spelling
    _____ Correct punctuation
    _____ Sentence structure
    _____ Paragraph structure
    _____ Other: _____
    _____ _____

4. Before taking this course, had you experienced:

A peer edit from a classmate?	Yes: _____	No: _____
A peer (colleague) edit at work?	Yes: _____	No: _____
An edit by a non-peer (not teacher)?	Yes: _____	No: _____

5. How did you feel about the experience of having your work edited by a classmate? Give examples to support response.

6. How did you feel about the experience of editing your classmate's writing? Give examples to support response.

[Use the back of this sheet, if necessary.]

APPENDIX B:
Samples from Questionnaires Distributed in Organizational Settings

Selected Quantitative Questions

General Information:
Questions covering demographics, such as job title and function, type of industry, organization of publication group, types of documents produced, kinds of editing performed and who performs them, and the (non)existence of written editorial standards.

Reactions to Editing:
Questions to which participants were asked to respond in terms of some sort of ranking scale, such as: strongly agree, agree, indifferent, disagree, or strongly disagree. Sample questions:

(a) Procedural and Quality Issues:

1. In most cases, a professional editor is more effective than a peer reviewer in overall document improvement.

2. Professional editors prefer working on documents that have received peer reviews.

3. The quality of documents improves if there is a peer review before the work is given to a professional editor.

4. I have the chance to talk about writing and/or editing strategies with my peers.

5. Peer reviewing improves a document in the following ways: [list follows for ranking of same components as on student questionnaire].

(b) Psychological Issues:

1. Peer editing motivates writers to invest more thought and energy into the rewrite phase of the job.

2. Personality differences affect my ability to be objective when editing another writer's work.

3. Writers always resist giving highly critical feedback to one another.

(c) Management Issues:

1. Managing the peer review process is best accomplished among peers.

2. Managing editors should actively direct peer editing, assigning the appropriate people to work together.

3. Managing the peer review process is best accomplished when the editorial guidelines are formally set by a supervisor.

Selected Qualitative (Open-Ended) Questions

On the Respondent's Environment:

1. Describe briefly the most significant characteristics of your working (writing and/or editing) environment.

2. What are the major strengths/weaknesses of your current editing system?

3. What changes would you make in your current editing system?

4. What kind of direction, if any, do you get?

5. Does your organization have formal editing procedures? If so, why do you think it adopted these procedures?

On Views of Peer and Hierarchical Editing:

1. Which type of editing do you think is more valuable (peer or hierarchical), and why?

2. Which type of editing do you prefer (peer or hierarchical), and why?

3. What are the major strengths/weaknesses of your current editing system?

4. What changes would you make in your current editing system?

5. If you have worked as a writer (editor) in the past, what did you like best/ least about your job?

6. Have you ever experienced an awkward working situation as a result of peer editing? If yes, describe this situation.

7. Do you ever worry that your peer relationships may be adversely affected by the same peers who edit your work? Explain why or why not.

On Training Requirements:

1. What did you have to learn to do your job effectively?

2. In your opinion, what skills should be taught in both industry and academia to ensure the success of editors who are employed in industry?

3. What sort of courses, workshops, or seminars have you found most useful to you as a writer? as an editor? What kinds of courses, etc., would you *like* to take?

4. How can your job develop you to take on more responsibility?

REFERENCES

1. L. M. Zook, Technical Editors Look at Technical Editing, *Technical Communication, 30*:3, pp. 20–26, 1983.
2. P. Elbow, *Writing Without Teachers*, Oxford University Press, London, 1973.

3. C. D. Rude (ed.), *Teaching Technical Editing,* Anthology Number 6, The Association of Teachers of Technical Writing, Lubbock, Texas, 1985.

4. C. J. Forbes, Developing Editing Skills in the Beginning Technical Writing Class, *The Technical Writing Teacher, 13*:2, pp. 122-134, 1986.

5. K. Bruffee, Collaborative Learning and the "Conversation of Mankind," *College English, 46*:7, pp. 635-652, 1984.

6. _____, Collaborative Learning: Some Practical Models, *College English, 34*:5, pp. 634-643, 1973.

7. R. W. Norton, Foundation of Communicator Style Construct, *Human Communication Research, 4,* pp. 99-112, 1978.

8. C. W. Downs, J. Archer, J. McGrath, and J. Stafford, An Analysis of Communication Style Instrumentation, *Management Communication Quarterly, 1*:4, pp. 543-571, 1988.

9. L. S. Chavarria, Using Workshop Sessions in Teaching Technical Writing, *The Technical Writing Teacher, 9,* pp. 95-99, 1982.

Selected, Annotated Bibliography on Collaborative Writing

MARGARET BATSCHELET
WILLIAM M. KARIS
THOMAS TRZYNA

INTRODUCTION

This bibliography is divided into four sections, each of which is intended to offer an introductory guide to published work on collaboration and its relationship to one of four key topics or issues: collaboration in the workplace, collaboration in the classroom, collaboration and technology, and collaboration and composition theory. Given the range of topics covered by chapters in this volume, we believe the relevance of these four issues to collaboration is obvious. Additionally, the relationships among these issues ought to be seen as symbiotic.

Readers should note that we have not attempted to incorporate the chapters contained within this volume or to duplicate all of the citations which may appear in the references to the individual chapters contained in this volume, though there is considerable overlap nonetheless. The bibliography contains one item from as far back as 1970, though the bulk of the citations are from between 1982-1987, the time when academe discovered the importance of collaborative writing.

Each section is prefaced with a short survey of the works contained in that section. The surveys attempt to clarify which works relate to particular sub-topics within that section's topic.

COLLABORATION IN THE WORKPLACE

Studies of collaboration in the workplace can be roughly divided into those suggesting procedures to be followed in order to improve the process of collaboration and those presenting research results based on studies of professional writing. There is also a small group of articles on science writing.

Techniques for improving the activities of writing teams are discussed by a number of articles. Abshire and Culbertson, Dressel (1983a), Fowler and Roger, Henkel, Hoerter and Brenner, Knapp, J. Mann, McAuslan and Armstrong, and Tracey concern organizing the team, while those by Hulbert, Mai, and Stoner all deal with managing it. Recommended procedures for accomplishing a team writing task are described by Dressel (1983b), Fridie, Minerly, Moorhead, Patton, Paxton, C. Smith, and Turner. Team testing of documents is discussed by Bagby.

Other relationships are also considered under the category of improving collaborative procedures. Adams, Boston, and P. Smith deal with team editing, while M. Mann, Moy, B. Myers, Simpson, Sims, and Thompson concern improvements in the writer/editor relationship. Palokoff deals with the collaboration between writer and artist, while Fritz considers assigning author credit in scientific and medical group work.

Research reports can be divided into survey results and analytical studies of professional writing. In the category of surveys, the extensive studies by Ede and Lunsford represent the most detailed surveys of collaborative writing in industry. The surveys by Anderson and by Faigley and Miller concern other topics as well, but both present some material on collaboration.

Analytical studies of professional writing which also present some material on collaboration include those by Broadhead and Freed, Debs, Doheny-Farina (1984), Flatley, G. Myers (1985), Odell, Paradis, and Selzer.

Research results are applied to collaborative work in the classroom by Doheny-Farina (1985) and Lutz, while Fridie and Watson apply workplace experiences to the classroom.

Finally, bibliographical studies of science writing, emphasizing the sociology of science and collaboration, are presented by Bazerman and by Myers (1986). The effect of collaboration on scientific work is considered by Myers (1985) and by Weinberg.

Abshire, G., Peer Reviews Improve Your Writing, *ACM SIGDOC Newsletter 9*:2, p. 4, 1983.

Abshire, G. and D. Culbertson, A Team Approach to Producing Good Documentation, *IEEE Transactions PC, 28*:4, pp. 38–41, 1985.

Adams, M., ". . . And Whether Pigs Have Wings:" Working with an Editorial Committee, *Editorial Eye, 123*, pp. 1–2, 1985.

Allen, J., The Mentor-Advisee Relationship, *Proceedings 32nd ITCC* (Houston), STC, Washington, D. C., MPD pp. 19–20, 1985.

Allen, N., D. Atkinson, M. Morgan, T. Moore, and C. Snow, What Experienced Collaborators Say about Collaborative Writing, *Iowa Journal of Business and Technical Communication, 1*:2, pp. 70-90, 1987.

Anderson, P., What Survey Research Tells Us about Writing at Work, in *Writing in Nonacademic Settings,* Odell and Goswami (eds.), The Guilford Press, New York, pp. 3-83, 1985.

Bagby, S., Improving Documents with Multiple Field-Test Groups, *Proceedings 31st ITCC* (Seattle), STC, Washington, D. C., RET pp. 58-60, 1984.

Baim, J., In-House Training in Report Writing: A Collaborative Approach, *ABC Bulletin, 40*:4, pp. 5-8, 1977.

Bazerman, C., Scientific Writing as a Social Act: A Review of the Literature of the Sociology of Science, in *New Essays in Technical and Scientific Communication: Research, Theory, and Practice,* Anderson, Brockman, and Miller (eds.), Baywood Publishing Company, Inc., Amityville, New York, pp. 156-184, 1985.

Bishop, C., Peer Review of Journal Articles in the Sciences, *Scholarly Communication around the World: Proceedings of Joint Global Conference,* SSP, Philadelphia, pp. 10-12, 1983.

Boston, B., Team Editing, *Editorial Eye, 134,* pp. 7-8, 1986.

Brinkman, K., The Team Multilingual Terminology Bank, *Technical Communication, 29*:4, pp. 6-7, 1982.

Broadhead, G. and R. Freed, *The Variables of Composition: Process and Product in a Business Setting,* Southern Illinois University Press, Carbondale, 1986.

Chadwick, K., The Professional Communications Network: A Research Process for Technical Writers, *The Technical Writing Teacher, 11*:3, pp. 210-220, 1984.

Debs, M., Collaborative Writing: A Study of Technical Writing in the Computer Industry, *DAI, 47*:6, p. 2141, 1986.

Dillon, L., Three Approaches to Writing for Group Acceptance, *The Technical Writing Teacher, 11*:3, pp. 186-189, 1984.

Doheny-Farina, S., A Mission to Collaborate: What Recent Ethnographic Research into Collaborative Writing Tells Teachers of Technical and Business Writing, *Proceedings of 32nd ITCC* (Houston), STC, Washington, D. C., RET pp. 56-59, 1985.

_____, Writing in an Emerging Organization: An Ethnographic Study, *Written Communication, 3*:2, pp. 158-185, 1986. Also Urbana, Illinois: ERIC/RCS and NCTE, 1984.

Dressel, S. et al., ASAPP: Automated Systems Approach to Proposal Production, *IEEE Transactions PC, 26*:2, pp. 63-68, 1983a.

Dressel, S., Generating a Quick First Draft, *IEEE Transactions PC, 26*:4, pp. 172-174, 1983b.

Ede, L. and A. Lunsford, Collaboration in Writing on the Job: A Research Report, *RIE, 21*:9, pp. 60-61, 1986. ED 268-582.

_____, Collaborative Learning: Lessons from the World of Work, *WPA, 9*:3, pp. 17-26, 1986.

_____, *Collaborative Writing: The Theory and Practice of Group Authorship,* Southern Illinois University Press, Carbondale (forthcoming).

Ede, L. and A. Lunsford, Research into Collaborative Writing, *Technical Communication, 32*:4, pp. 69–70, 1985.

_____, Research on Co- and Group Authorship in the Professions: A Preliminary Report, *RIE, 20*:10, p. 36, 1985. ED 257086.

_____, Why Write ... Together: A Research Update, *Rhetoric Review, 5*:1, pp. 71–77, 1986.

Faigley, L. and T. Miller, What We Learn from Writing on the Job, *College English, 44*:6, pp. 557–569, 1982.

Flatley, M., A Comparative Analysis of the Written Communication of Managers of Various Organizational Levels in the Private Business Sector, *Journal of Business Communication, 19*:3, pp. 35–49, 1982.

Fowler, S. and D. Roger, Programmer and Writer Collaboration: Making User Manuals That Work, *IEEE Transactions PC, 19*:4, pp. 21–25, 1986.

Fridie, P., Interpersonal Skills for Technical Writers, *The Technical Writing Teacher, 13*:3, pp. 316–317, 1986.

Fritz, J., Who Gets Credit As Author, *Medical Communications, 13*:1, pp. 1–4, 1985.

Gerkin, P., The Production Manager: From Artist to Traffic Cop, *Editorial Eye, 125*, pp. 4–5, 1986.

Hartley, J., The Role of Colleagues and Text-Editing Programs in Improving Texts, *IEEE Transactions PC, 27*:1, pp. 42–44, 1984.

Henkel, G., Problems of Communication in Project Planning and Development, *Journal of Technical Writing and Communication, 11*:1, pp. 9–12, 1981.

Hoerter, G. and T. Brenner, Teamwork ... The Key to Managing Proposal Production, *Proceedings 29th ITCC* (Boston), STC, Washington, D. C., C pp. 45–48, 1982.

Hulbert, J., Conducting Intelligent Business Dialog, *IEEE Transactions PC, 24*:4, pp. 132–135, 1981.

Knapp, J., A New Role for the Technical Communicator: Member of a Design Team, *Proceedings 31st ITCC* (Seattle), STC, Washington, D. C., WE pp. 30–33, 1984.

Lutz, J., The Influence of Organizations on Writers' Texts and Training, *The Technical Writing Teacher, 13*:2, pp. 187–190, 1986.

Mai, M., Using Project Management to Improve Communications, *Proceedings 33rd ITCC* (Detroit), STC, Washington, D. C., pp. 125–128, 1986.

Mann, J., The Engineer/Writer Team Approach in Preparing a Technical Paper, *Proceedings 30th ITCC* (St. Louis), STC, Washington, D. C., pp. 7–9, 1983. Also in ERIC ED 239 275; *RIE, 19*:6, p. 42, 1984.

Mann, M., How to Edit the Passive Writer's Work, *Technical Communication, 32*:3, pp. 14–15, 1985.

McAuslan, G. and F. Armstrong, Using a Team Approach for Better Documentation, *Proceedings 31st ITCC* (Seattle), STC, Washington, D. C., VC pp. 36–39, 1984.

Minerly, K., Peer Education: Sharing the Skills in a Team Environment, *Proceedings 30th ITCC* (St. Louis), STC, Washington, D. C., MPD pp. 7–8, 1983.

Moorhead, A., The Conference Approach to Engineers' Report Writing, *IEEE Transactions PC, 28*:3, pp. 13–16, 1985.

Moy, C., Editing the Prima Donna, *Editorial Eye, 120,* pp. 1–3, 1985.

Myers, B., A Classification of Author-Editor Relationships: Toward Team-Centered Relationships, *Proceedings 31st ITCC* (Seattle), STC, Washington, D. C., WE pp. 116–119, 1984.

Myers, G., The Social Construction of Two Biologists' Proposals, *Written Communication, 2*:3, pp. 219–245, 1985.

_____, Writing Research and the Sociology of Scientific Knowledge: A Review of Three New Books, *College English, 48*:6, pp. 595–610, 1986.

Odell, L., Beyond the Text: Relations Between Writing and Social Context, in *Writing in Nonacademic Settings,* Odell and Goswami (eds.), The Guilford Press, New York, pp. 249–280, 1985.

Palokoff, K., How Management Decisions Affect Writer-Artist Collaboration, *Proceedings 32nd ITCC* (Houston), STC, Washington, D. C., MPD pp. 32–34, 1985.

Paradis, J., D. Dobrin, and R. Miller, Writing at Exxon ITD: Notes on the Writing Environment of an R&D Organization, in *Writing in Nonacademic Settings,* Odell and Goswami (eds.), The Guilford Press, New York, pp. 281–307, 1985.

Patton, J., The Quest for Clout and Credibility: How a Publishing Group Can Survive in a High-Tech Environment, *Proceedings 31st ITCC* (Seattle), STC, Washington, D. C., MPD pp. 75–78, 1984.

Paxton, A., How an Experienced Writer Can Help a Novice, *Proceedings 32nd ITCC* (Houston), STC, Washington, D. C., MPD pp. 24–27, 1985.

Polanski, V., JFK and Team: Effective Style Through Parallel Structure, *Technical Communication, 32*:2, p. 64, 1985.

Selzer, J., The Composing Process of an Engineer, *College Composition and Communication, 34*:2, pp. 178–187, 1983.

Simpson, A., Editors and Usability Testing: A Perfect Match, *Proceedings 32nd ITCC* (Houston), STC, Washington, D. C., MPD pp. 65–67, 1985.

Sims, R., Advantages of Dialogue from a Management Perspective, *Proceedings 31st ITCC* (Seattle), STC, Washington, D. C., W&E pp. 114–115, 1984.

_____, Dialogue: The Key to Professionalism in Technical Communications, *Proceedings 30th ITCC* (St. Louis), STC, Washington, D. C., pp. 35–37, 1983. Also ED 239 278; *RIE, 19*:6, pp. 43–44, 1984.

Smith, C., Word Processing and Scientific Writing in a University Research Group, *Technical Communication, 29*:3, pp. 13–17, 1982.

Smith, P., Proofreading: Solo or Team?, *Editorial Eye, 101,* pp. 1–2, 1984.

Soderson, C., The Usability Edit: A New Level, *Technical Communication, 32*:1, pp. 16–18, 1985.

Stoner, R., Conflict Management for the Communications Supervisor, *Proceedings 32nd ITCC* (Houston), STC, Washington, D. C., MPD pp. 58–60, 1985.

Thompson, M., Easy Does It: Author and Author's Editor, *Proceedings 32nd ITCC* (Houston), STC, Washington, D. C., p. 64, 1985.

Tracey, J., The Theory and Lessons of STOP Discourse, *IEEE Transactions PC, 26*:2, pp. 68–78, 1983.

Turner, G., Professionalism and Time Management: A Look at Structured Cooperation in a Technical Workgroup, *Proceedings 32nd ITCC* (Houston), STC, Washington, D. C., MPD pp. 35–38, 1985.

Watson, R., Field-Paradigms: The Relation between the Technical Communicator and the Technical Team, *Proceedings 1984 CPTSC*, CPTSC, Santa Fe, pp. 159–179, 1984.

Weinburg, A., Scientific Teams and Scientific Laboratories, *Daedalus, 99*:4, pp. 1056–1075, 1970.

COLLABORATION IN THE CLASSROOM

Goldstein (1984) presents a general review of methods for using collaboration in the classroom.

The preponderance of the pedagogical articles on collaboration address procedures and questions for organizing peer review. These essays include those by Baxter and Clark, Chavarria, Cullinan, Doheny-Farina, Flynn et al., Forbes, Hageman and Vest, Lauerman, Myers, Norman and Young, and Sills. Collins focuses on peer review in business writing courses, Duin on guidelines for "cooperative learning groups," and Jacko on the roles taken by students who are working in groups. Within the general subject of peer review, a sub-group of essays discusses the role of talk, dialogue, or monologue in group dynamics. Gere and Abbott evaluate talk, while David and Goldstein (1981) investigate dialogue. Kraft describes a method that involves oral reports, and Pufahl examines the talk of developmental writers.

Joint authorship and/or team writing are examined by Adams, Early and Statz, Kelly, and Tebeaux. Daiute focuses on joint authorship among school children.

Group management is more particularly the focus of essays by Baxter and Clark, and O'Donnell et al. Hawkins offers a long and complete examination of group protocols, while Smith addresses the interpersonal skills necessary for good group process, and Walvoord describes a sequential model for organizing a group's work.

Approaches to evaluating collaborations are described by Baxter and Clark, Gere and Abbott, and Gross. Newkirk examines the disparity between instructors' and students' evaluations of collaborative groups. George suggests methods for helping groups that are not working effectively, while Doheny-Farina identifies problems that can undermine collaboration. Lutz presents a study of the organizational or socio/political constraints on collaboration in industry.

Topics for collaborative exercises are presented by Brown, Burnham, Gross, and Potvin. This bibliography does not include the collaborative exercises included in recent textbooks or casebooks. Several articles focus on other aspects of collaboration: Covington on oral and visual reports, two essays by Goldstein and Malone (1984, 1985) on the value of journals to collaboration, and Fisher and Atkinson on the connection between collaboration and brainstorming or invention.

At a more theoretical level, Forman and Katsky, Hord, Levy-Reiner, Poole et al., and Reither explore the distinctions among small group theory, collaboration, writing process theory, and models of cooperation.

Adams, E., Joint Authorship: A Pedagogical Methodology to Increase Awareness of Audience, *DAI, 46*:8, p. 2213A, 1986.

Addams, H., Accountants and Communications Instructor Team to Upgrade Writing of Accounting Majors, *Proceedings 1984 ABCA,* Neal, Scott, and Lundgren (eds.), ABCA, Salt Lake City, pp. 57–60, 1984.

Baxter, C. and T. Clark, My Favorite Assignment: Putting Organization and Interpersonal Communication Theory into Practice: Classroom Committees, *ABCA Bulletin, 45*:3, pp. 38–41, 1982.

Brown, S., More than an Exercise: Annotated Bibliography as Collaborative Learning, *RIE, 22*:6, p. 35, 1987. Ed 278 018.

Brumback, T., Jr., Peer Response: An Effective Way to Incorporate Writing into the Classroom, *NACTA Journal, 29*:1, pp. 77–81, 1985.

Burnham, C., The Consequences of Collaboration: Discovering Expressive Writing in the Disciplines, *The Writing Instructor, 6*:1, pp. 17–24, 1986.

Chavarria, L., Using Workshop Sessions in Teaching Technical Writing, *The Technical Writing Teacher, 9*:2, pp. 95–99, 1982.

Collaborative Learning, Levy-Reiner (ed.), *RIE, 21*:4, p. 89, 1986. ED 263 868.

Collins, T., What's New in Freshman Comp? Some Gleanings for the Business Writing Teacher, *ABCA Bulletin, 44*:1, pp. 10–12, 1981.

Covington, D., Making Team Projects Work in Technical Communications Courses, *The Technical Writing Teacher, 11*:2, pp. 167–174, 1984.

Cullinan, M., Developing Business Writing Skills Through Group Activity, *ABCA Bulletin, 50*:1, pp. 21–22, 1987.

Daiute, C., Do 1 and 1 Make 2? Patterns of Influence of Collaborative Writers, *Written Communication, 3*:3, pp. 382–408, 1986.

David, D., An Ethnographic Investigation of Talk in Small Group Writing Workshops in a College Writing Class, *DAI, 17*:6, p. 1061, 1986.

Doheny-Farina, S., A Mission to Collaborate: What Recent Ethnographic Research into Collaborative Writing Tells Teachers of Technical and Business Writing, *Proceedings of 32nd ITCC* (Houston), STC, Washington, D. C., RET pp. 56–59, 1985.

Duin, A., Implementing Cooperative Learning Groups in the Writing Curriculum: What Research Shows, Paper presented at the Minnesota Council of Teachers of English Meeting, Mankato, Minnesota, May 1984. ERIC ED 251 849 [17 pages].

Early, D. and C. Stutz, Teaching "Team" Research Techniques and Technical Report Writing in Elementary Physics Laboratories, *American Journal of Physics, 44*:10, pp. 953–955, 1976.

Fisher, J. and T. Atkinson, Writing Surveys for Institutional Research: A Profitable Collaboration, *ABCA Bulletin, 46*:2, pp. 34–37, 1983.

Flynn, E. et al., Effects of Peer Critiquing and Model Analysis on the Quality of Biology Student Laboratory Reports, *RIE, 19*:2, p. 44, 1984. ED 234 403.

Forbes, C., Developing Editing Skills in the Beginning Technical Writing Class, *The Technical Writing Teacher, 13*:2, pp. 122–134, 1986.

Forman, J. and P. Katsky, The Group Report: A Problem in Small Group or Writing Processes?, *Journal of Business Communication, 23*:4, pp. 23–35, 1986.

George, D., Working with Peer Groups in the Composition Classroom, *College Composition and Communication, 35*:3, pp. 320–326, 1984.

Gere, A. and R. Abbott, Talking About Writing: The Language of Writing Groups, *Research in the Teaching of English, 19*:4, pp. 362–381, 1985.

Gerson, S., Peer Assistance, Tutorial Workshops, Instructional Modules: Help for Non-Native Students, *Proceedings 29th ITCC* (Boston), STC, Washington, D. C., E pp. 49–51, 1982.

Goldstein, J., Dialogue: The Critical Oral Skill for Students in Technical Writing, *The Technical Writing Teacher, 8*:3, pp. 54–58, 1981.

———, Trends in Teaching Technical Writing, *Technical Communication, 31*:4, pp. 25–34, 1984.

Goldstein, J. and E. Malone, Journals on Interpersonal and Group Communication: Facilitating Technical Project Groups, *Journal of Technical Writing and Communication, 14*:2, pp. 113–131, 1984.

———, Using Journals to Strengthen Collaborative Writing, *ABCA Bulletin, 48*:3, pp. 24–28, 1985.

Grimm, N., Classroom Management: Using Small Groups Effectively, *Arizona English Bulletin, 28*:1, pp. 52–60, 1985.

Gross, G., Group Projects in the Technical Writing Course, in *Courses, Components, and Exercises in Technical Communications,* Stephenson et al. (eds.), NCTE, Urbana, pp. 54–64, 1981.

Hageman, M. and L. Vest, A Monologue-Dialogue Workshop for Teaching Editing, *Technical Communication, 32*:4, p. 66, 1985.

Hawkins, T., *Group Inquiry Techniques for Teaching Writing,* NCTE and ERIC/RCS, Urbana, 1976. ED 128 813.

Hord, S., Collaboration or Cooperation: Comparisons and Contrasts, Dilemmas and Decisions, *RIE, 20*:11, p. 67, 1985. ED 258 356.

Jacko, C., Small-Group Triad: An Instructional Mode for the Teaching of Writing, *College Composition and Communication, 29*:3, pp. 290–292, 1978.

Jaffe, G., Teaching Collaborative Technical Writing Projects, *Proceedings 1985 CPTSC,* Marilyn Schauer Samuels (ed.), CPTSC, Oxford, Ohio, pp. 104–111, 1985.

Jordan, M., The Effects of Cooperative Peer Review on College Students Enrolled in Required Advanced Technical Writing Courses, *DAI, 45*:5, p. 1319A, 1984.

Kelly, P., M. Hall, and R. Small, Jr., Composition through the Team Approach, *English Journal, 73*:5, pp. 71–74, 1984.

Kraft, R., Group-Inquiry Turns Passive Students Active, *College Teaching, 33*:4, pp. 149–154, 1985.

Lauerman, D. et al., Workplace and Classroom: Principles for Designing Writing

Courses, in *Writing in Nonacademic Settings,* Odell and Goswami (eds.), The Guilford Press, New York, pp. 427–450, 1985.

Liggett, S., Collaborative Writing from Start to Finish, *Louisiana English Journal, 24*:2, pp. 24–29, 1985.

Lutz, J., The Influence of Organizations on Writers' Texts and Training, *The Technical Writing Teacher, 13*:2, pp. 187–190, 1986.

Myers, G., The Writing Seminar: Broadening Peer Collaboration in Freshman English, *The Writing Instructor, 6*:1, pp. 48–56, 1986.

Nees-Hatlen, V., Collaborating on Writing Assignments: A Workshop with Theoretical Implications, *Journal of Teaching Writing, 4*:2, pp. 234–246, 1985.

Newkirk, T., Direction and Misdirection in Peer Response, *College Composition and Communication, 35*:3, pp. 301–311, 1984.

Norman, R. and M. Young, Using Peer Review to Teach Proposal Writing, *The Technical Writing Teacher, 12*:1, pp. 1–9, 1985.

O'Donnell, A. et al., Cooperative Writing: Direct Effects and Transfer, *Written Communication, 2*:3, pp. 307–315, 1985.

Poole, M., D. Seibold, and R. McPhee, Group Decision Making as a Structurational Process, *Quarterly Journal of Speech, 71*:1, pp. 74–102, 1985.

Potvin, J., Using Team Reporting Projects to Teach Concepts of Audience and Written, Oral, and Interpersonal Communication Skills, *IEEE Transactions PC, 27*:3, pp. 130–137, 1984.

Pritchard, R., Who Needs the Practice? Research on Peer Writing Groups, *North Carolina English Teacher, 42*:1, pp. 22–23, 1984.

Pufahl, J., Five Writers Talking: A Case Study of the Collaborative Writing Conference, *DAI, 44*:2, p. 413, 1983.

Reither, J., What Do We Mean by "Collaborative Writing" (And What Difference Might It Make)?, *RIE, 22*:8, p. 50, 1987. ED 280 084.

Rodrigues, D., Peer Review Possibilities, *Connecticut English Journal, 15*:2, p. 80, 1984.

Schuster, C., The Un-Assignment: Writing Groups for Advanced Expository Writers, *Freshman English News, 13*:3, pp. 4, 10–14, 1984.

Sharplin, W., The Cooperative Theme, *Louisiana English Journal, 23*:1, pp. 28–30, 1984.

Sills, C., The Scholar Who Helps Me Teach Better: Adapting Freewriting Techniques and Writing Support Groups for Business Communications, *ABCA Bulletin, 48*:2, pp. 12–14, 1985.

Smith, H., Methods for Training the Technical Editor in Interpersonal Skills, *IEEE Transactions PC, 28*:1, pp. 46–50, 1985.

Tebeaux, E., Redesigning Professional Writing Courses to Meet the Communication Needs of Writers in Business and Industry, *College Composition and Communication, 36*:4, pp. 419–428, 1985.

Tritt, M., Collaboration in Writing: From Start to Finish and Beyond, *English Quarterly, 17*:1, pp. 82–86, 1984.

Walvoord, B., Student Response Groups: Training for Autonomy, *The Writing Instructor, 6*:1, pp. 39–47, 1986.

COLLABORATION AND TECHNOLOGY

One productive area for research has been the effect of developing communications technology upon collaboration. Word processing, computer networks, computer conferences, electronic mail, and video conferences all can involve collaborative work to some degree.

Introductory material on technology with some reference to collaboration is provided by several selections. Caernarvan-Smith provides a basic introduction to network terminology, while both Feenberg and Penrose describe computer conferences, and Losey describes electronic mail. Penrose's article also provides an introduction to video conferences.

The relationship between word processing and collaboration is described by Arms, Halpern and Liggett, Hawisher and Schmidt, and Selfe and Wahlstrom. Arms (1985), Hawisher and Schmidt, and Selfe and Wahlstrom are concerned with computer use and collaboration in the classroom, while Arms (1983) and Halpern and Liggett discuss collaborative computer use in the workplace.

Electronic mail and its effect on group work are discussed by Losey, Steinfield, and Halpern and Liggett, while Rosetti and Surynt present the results of a study of the efficiency of video conferences.

Computer conferencing and collaboration concern the largest group of articles in this section, while those by Kerr, Pfaffenberger, and Spitzer are particularly concerned with group interaction in a computer conference.

Arms, V., The Computer: An Aid to Collaborative Writing, *The Technical Writing Teacher, 11*:3, pp. 181–185, 1985.

_____, Engineers Like to Write—On a Computer!, *IEEE Transactions PC, 26*:4, pp. 175–177, 1983.

Bell, A. and T. Housel, Teleconferencing Comes of Age—Again, *Technical Horizons in Education, 13*:9, pp. 71–81, 1986.

Caernarvan-Smith, P., Computers and Communication: Networks—The Newest Old Technology, *Technical Communication, 33*:1, pp. 41–43, 1986.

Caton, S., Local Area Networks and Their Impact on Technical Communications, *Proceedings 31st ITCC* (Seattle), STC, Washington, D.C., ATA pp. 83–85, 1984.

Feenberg, A., Network Design: An Operating Manual for Computer Conferencing, *IEEE Transactions PC, 29*:1, pp. 2–7, 1986.

Halpern, J. and S. Liggett, *Computers and Composing: How the New Technologies Are Changing Writing*, Southern Illinois University Press, Carbondale, 1984.

Hawisher, G. and G. Schmidt, Collaborative Writing: A Successful Strategy for Computer-Aided Instruction, *Illinois English Bulletin, 73*:1, pp. 28–35, 1985.

Kerr, E., Electronic Leadership: A Guide to Moderating Online Conferences, *IEEE Transactions PC, 29*:1, pp. 12–18, 1986.

Losey, C., Electronic Messaging Systems for More Effective Management, *IEEE Transactions PC, 28*:3, pp. 35–39, 1985.

Nightingale, J., Computer Conferencing: A New Form of Communication, *Proceedings 29th ITCC* (Boston), STC, Washington, D. C., T pp. 51-54, 1982.

Penrose, J., Telecommunications, Teleconferencing, and Business Communication, *Journal of Business Communication, 21*:1, pp. 93-111, 1984.

Pfaffenberger, B., Research Networks, Scientific Communication, and the Personal Computer, *IEEE Transactions PC, 29*:1, pp. 30-33, 1986.

Pullinger, D., Chit-Chat to Electronic Journals: Computer Conferencing Supports Scientific Communication, *IEEE Transactions PC, 29*:1, pp. 23-29, 1986.

Rosetti, D. and T. Surynt, Video Teleconferencing and Performance, *Journal of Business Communication, 22*:4, pp. 25-31, 1985.

Selfe, C. and B. Wahlstrom, An Emerging Rhetoric of Collaboration: Computers, Collaboration, and the Composing Process, *RIE, 21*:2, p. 40, 1986. ED 261 384.

Spitzer, M., Writing Style in Computer Conferences, *IEEE Transactions PC, 29*:1, pp. 19-22, 1986.

Steinfield, C., Communication via Electronic Mail: Patterns and Predictors of Use in Organizations, *DAI, 45*:1, p. 5A, 1984.

COLLABORATION AND COMPOSITION THEORY

It is likely that the prime motivation behind the increased attention given to collaborative activities over the past decade and a half has been technical communicators' awareness of how prevalent collaboration is in industry. Teachers of writing now perceive that knowledge is socially developed or constructed. This recognition helps focus attention on the theoretical underpinnings of collaboration [See, for example, Lipson]. The items in this section do not lend themselves as neatly to the type of categorization which our earlier surveys have presented. Consequently, the descriptions here attempt to be more evaluative as well as indicate some sense of the focus each item offers.

Bruffee (1983, 1984, 1986), Cooper, Faigley, Hairston, and Trimbur offer valuable theoretical introductions. The newcomer to collaboration may find Trimbur particularly accessible. Gere offers a broader historical perspective as well as an overview of theories of collaborative learning [see her chapter 3 especially]. Lefevre provides a valuable summary of theoretical perspectives, including discussion of the philosophy of knowledge which underlies the entire collaborative approach [see her chapter 5 especially].

Bruffee (1973), Clifford, and Gebhardt offer thoughtful advice on some pedagogical implications and outcomes of collaborative theory.

As collaboration becomes more fully the predominant pedagogical stance taken by a majority of writing teachers, minority objections to its primacy are beginning to be heard. Myers, for example, offers a thoughtful critique of collaboration's theoretical position. To date, perhaps Bizzell, Ede, and Maimon might be seen as "apologies" for the position, though only Ede's could correctly be termed a direct response [to Myers' *College English* essay].

Bizzell, P., Foundationalism and Anti-Foundationalism in Composition Studies, *PRE/TEXT,* 7:1–2, pp. 37–56, 1986.

Bruffee, K., Collaborative Learning: Some Practical Models, *College English,* 34:5, pp. 634–643, 1973.

_____, Collaborative Learning and the "Conversation of Mankind," *College English, 46*:7, pp. 635–652, 1984.

_____, Social Construction, Language, and the Authority of Knowledge: A Bibliographic Essay, *College English, 48*:8, pp. 773–790, 1986.

_____, Writing and Reading as Collaborative or Social Acts, in *The Writer's Mind,* Hays, Roth, Ramsey, and Foulke (eds.), NCTE, Urbana, 1983.

Clifford, J., Composing in Stages: The Effects of Collaborative Pedagogy, *Research in the Teaching of English, 15*:1, pp. 37–53, 1981.

Cooper, M., The Ecology of Writing, *College English, 48*:4, pp. 364–375, 1986.

Ede, L., The Case for Collaboration, *RIE, 22*:10, p. 39, 1987. ED 282 212.

Faigley, L., Competing Theories of Process: A Critique and a Proposal, *College English, 48*:6, pp. 527–542, 1986.

Gebhardt, R., Teamwork and Feedback: Broadening the Base of Collaborative Writing, *College English, 42*:1, pp. 69–74, 1980.

Gere, A., *Writing Groups: History, Theory, and Implications,* Southern Illinois University Press, Carbondale, 1987.

Hairston, M., The Winds of Change: Thomas Kuhn and the Revolution in the Teaching of Writing, *College Composition and Communication, 33*:1, pp. 76–88, 1982.

LeFevre, K., *Invention as a Social Act,* Southern Illinois University Press, Carbondale, 1987.

Lipson, C., A Social View of Technical Writing, *Iowa State Journal of Business and Technical Communication, 2*:1, pp. 7–20, 1988.

Maimon, E., Knowledge, Acknowledgement, and Writing Across the Curriculum: Toward an Educated Community, in *The Territory of Language: Linguistics, Stylistics, and the Teaching of Writing,* Donald McQuade (ed.), Southern Illinois University Press, Carbondale, pp. 89–100, 1986.

Myers, G., Reality, Consensus, and Reform in the Rhetoric of Composition Teaching, *College English, 48*:2, pp. 154–171, 1986.

Trimbur, J., Collaborative Learning and Teaching Writing, in *Perspectives on Research and Scholarship in Composition,* McClelland and Donovan (eds.), MLA, New York, pp. 87–109, 1985.

Index

Contributors

MARGARET BATSCHELET teaches English and technical writing at the University of Texas at San Antonio. She is the author, with Thomas Trzyna, of *Writing for the Technical Professions* and *The Technical Writing Casebook* (both Wadsworth). As a consultant she has written and edited various technical documents for Pipe Creek Publications, Information Transfer and Datapoint Corporation; she has also conducted training seminars for business and government groups.

BARBARA COUTURE is Associate Professor of English and Director of Composition at Wayne State University where she teaches graduate and undergraduate courses in technical writing, composition, and writing research and theory. She is contributor to and editor of *Functional Approaches to Writing: Research Perspectives* (Ablex, 1986) and *Professional Writing: Toward a College Curriculum* (ATTW, 1987); and co-author of *Cases for Technical and Professional Writing* (Little, Brown, 1985).

MARK DeBOWER is Research Systems Manager at Cargill, Inc. He has worked on numerous group projects with landscape architecture companies, and he regularly teaches classes on telecommunications. His research interests are in the areas of telecommunications and group work.

ANN HILL DUIN is Assistant Professor of Rhetoric at the University of Minnesota where she directs courses in technical communication. She recently won the EDUCOM/NCRIPTAL 1989 national award for the best curriculum innovation in the areas of computers and writing. Her research interests include the effects of computers and telecommunications on users' cognitive processes and group writing patterns.

DAVID K. FARKAS is an Associate Professor in the Department of Technical Communication, College of Engineering, University of Washington. His research interests include the advisory interface of computer systems, print and electronic document design, and hypermedia. His publications have received awards from the Society for Technical Communication (1982) and from the NCTE (1987 and 1989).

ALICE GILLAM is an Assistant Professor of English at the University of Wisconsin–Milwaukee where she teaches beginning and advanced writing and coordinates the peer tutoring program. Involved in peer tutoring programs since 1982, Gillam has published several articles on peer tutoring and collaborative learning. Currently, she is writing a book-length study of peer tutoring theory, research, and practice, entitled *Voices from the Center*.

ROGER A. GRICE is Advisory Information Developer at IBM's Kingston, New York, laboratory and Adjunct Professor at Rensselaer Polytechnic Institute. He has been a member of IBM's information-development organization for the past 23 years and is currently involved in large-systems documentation, information usability, and online information. He is a senior member of the Society for Technical Communication and is Director-Sponsor for Region 1, manager of the nominating committee, and manager of the scholarships subcommittee. He is also a senior member of the Institute of Electrical and Electronics Engineers and a member of the Advisory Committee Institute's Professional Communication Society.

DIXIE ELISE HICKMAN, President of the Association of Professional Writing Consultants, has edited *Resources for Writing Consultants* and *Teaching Technical Writing: Graphics*. A university teacher for 17 years and writing program administrator for 5, she is now owner of Communication Resources, which provides communication services and training.

LINDA A. JORN is the Instructional Computing Coordinator for the College of Agriculture at the University of Minnesota where she is pursuing a Masters of Science Degree in Technical Communication. Prior to her current work, she was a registered nurse for ten years. Her research interests are in the areas of computer-assisted instruction and collaboration.

WILLIAM M. KARIS is Associate Professor of Technical Communications at Clarkson University. He has published in such journals as *Rhetoric Review, Technical Communication,* and the *Technical Writing Teacher.* He was a participant in the 1987 NEH Summer Seminar on Rhetoric, held at Ohio State under the direction of Edward P. J. Corbett. His current research interests are collaborative writing and environmental rhetoric.

MARY M. LAY is Associate Professor and Chairperson of Technical Communications at Clarkson University. She was the 1988–1990 President of the Association of Teachers of Technical Writing and author of *Strategies for Technical Writing* (Holt, Rinehart and Winston, 1982). She has published articles in such journals as *The Technical Writing Teacher,* the *Journal of Business and Technical Communication,* and *Technical Communication.* Her research interests are collaboration and gender.

ELIZABETH L. MALONE is a Technical Writer for General Sciences Corporation, a company primarily engaged in designing and supporting computer systems; she works in the Contracts and Grants Division of NASA Headquarters, in collaboration with many different NASA program managers and contract specialists. She previously taught technical writing at Wayne State University and participated in that university's federally funded "Professional Writing Project." Malone has written and co-authored articles on using journals to facilitate small project groups, several communications cases, and *Professional Writing: Toward a College Curriculum.*

MEG MORGAN is Assistant Professor of English at the University of North Carolina at Charlotte. She teaches graduate courses in Rhetoric and Composition theory, technical communication, advanced and freshman composition. She has published in *The Iowa State Journal of Business and Technical Communication* and in the *Bulletin of the Association for Business Communication.* She is an associate editor for *Technical Communication.*

MARY MURRAY is Assistant Professor in the Rhetoric Department at Hobart and William Smith Colleges in Geneva, NY. Her current research interests include the measurement of writing, writing program evaluation, and the history of grammar.

JONE RYMER is Associate Professor and Coordinator of Business Communication, School of Business Administration, Wayne State University. Her recent publications include contributions to *Writing in Academic Disciplines* (Ablex, 1988), *Writing in the Business Professions* (NCTE/ABC, 1989), the *Journal of Business and Technical Communication* (1989), and *Collaborative Technical Writing: Theory and Practice* (ATTW, 1989). She is co-author of *Cases for Technical and Professional Writing* (Little, Brown, 1985).

HENRIETTA NICKELS SHIRK is Assistant Professor of English at Northeastern University. She teaches a variety of graduate-level courses in technical and professional writing and is Coordinator for Northeastern's Computer Writing Laboratory and Assistant Coordinator for the Technical Writing Training Program. Her articles have been published in *The Journal of Technical Writing and Communication, Data Training, Text, ConText and HyperText* (MIT Press), and *Collaborative Technical Writing: Theory and Practice* (ATTW). She is editor for *The Bulletin* of the Association for Business Communication.

ELIZABETH TEBEAUX is Chair of the CCCC Committee on Technical and Scientific Communication and Vice President of the Association of Teachers of Technical Writing (1990–1992). She has authored articles on style in both literature and technical communication, pedagogy, and trends in technical communication; *Writing Communication in Business and Industry* (Prentice-Hall, 1982); and *Design in Business Communication: The Process and the Product* (Macmillan, 1990). She is Coordinator of Technical Writing at Texas A&M University.

THOMAS TRZYNA is co-author of *Writing for the Technical Professions* (Wadsworth, 1987), which includes a chapter on collaborative writing, and *The Technical Writing Casebook* (also Wadsworth, 1987). His essays on composition have appeared in *College Composition and Communication* and *The Writing Instructor.* He is currently Dean of the School of Humanities at Seattle Pacific University.

WILLIAM VAN PELT is an Associate Professor of English at the University of Wisconsin-Milwaukee where he teaches composition, technical writing, and "Writing for Computer Technology." He helped establish the University's microcomputer writing lab and has coordinated the lab since 1984. Before coming to UWM, Professor Van Pelt worked as a technical writer and training consultant for several large corporations. He has published many articles on technical writing and writing with computers. He is currently writing a book entitled *The Rhetoric of Technical Discourse: Ethos, Pathos, and Logos in the Modern Age.*

JAMES WEBER has taught technical communication in the Scientific and Technical Communications Program at the University of Washington. He has also been a staff writer and consultant for the ERIC Clearinghouse on Educational Management, writing analyses of current research in education. Currently, he is a technical editor at Battelle Northwest, Richland, Washington, where he also teaches in the Battelle Writing Workshop program. His research interests include communication theory and education.

TIMOTHY WEISS is Assistant Professor of English at the University of Illinois. He has published articles on professional writing and a book on Ford Madox Ford. During 1988-89 he was a Fulbright lecturer in Tunisia, and is writing a critical study of V. S. Naipaul entitled *On the Margins: The Art of Exile.*